Silas Whitcomb Holman

Computation Rules and Logarithms

With Tables of Other Useful Functions

Silas Whitcomb Holman

Computation Rules and Logarithms
With Tables of Other Useful Functions

ISBN/EAN: 9783337130619

Printed in Europe, USA, Canada, Australia, Japan

Cover: Foto ©berggeist007 / pixelio.de

More available books at **www.hansebooks.com**

COMPUTATION RULES

AND

LOGARITHMS

WITH TABLES OF OTHER USEFUL FUNCTIONS

BY

SILAS W. HOLMAN

PROFESSOR OF PHYSICS AT THE
MASSACHUSETTS INSTITUTE OF TECHNOLOGY

New York
THE MACMILLAN COMPANY
LONDON: MACMILLAN & CO., LTD.
1896

Set up and electrotyped December, 1895. Reprinted May, October, 1896.

Norwood Press
J. S. Cushing & Co. – Berwick & Smith
Norwood Mass. U.S.A.

CONTENTS.

TABLES.

PREFACE.

It would probably be within safe limits to assert that one-half of the time expended in computations is wasted through the use of an excessive number of places of figures, and through failure to employ logarithms. This waste might be almost wholly avoided by following a few simple computation rules and practising slightly with logarithm tables.

The loss from the use of superfluous figures will be appreciated when it is considered that in direct or logarithmic multiplication and division with four, five, and six places of figures the work is respectively in the ratio of 1 : 2 : 3, or perhaps more nearly 2 : 3 : 4. Thus contrary to the fallacious excuse so commonly given that it is just about as easy to use six- or seven-place tables as smaller ones, the work is doubled or trebled by the use of six places instead of four. Even the employment of six- or seven-place tables, and dropping superfluous places when four or five are desired, causes much loss of time.

The proper employment of logarithms for work of four or more places effects a saving of one-quarter and upward of the time required for direct multiplication or division, with a lessening of fatigue and a gain of accuracy.

The following pages contain simple rules to enable one to answer for himself the question, how many places of figures ought I to use in this computation? — also, an explanation of the use of the notation by powers of ten; certain instructions, more or less novel in form, as to the use of the logarithm and other tables; and a collection of useful tables. This collection is designed to contain all the mathematical tables ordinarily required, and nothing more, in practical work in all branches of the engineering professions, and by students of physics, chemistry, and engineering, for work of any grade not exceeding about one-twentieth of one per cent in accuracy. For many persons the present volume should, therefore, provide all the logarithmic and trigonometric tables needed for the entire range of their practice. For work of greater precision than the above limit, the more bulky Vega, or some similar reliable seven-place table would be required. It is exceedingly rare that more than six or seven places are necessary, while for most work five are sufficient, although a striking chapter of absurd illustrations might be gleaned from various text-books and tables where ten- and even twenty-place logarithms are given, often for quantities uncertain in their fourth or fifth place. Persons doing much work with squares, cubes, square roots, cube roots, or reciprocals of more than four places would naturally make use of the Barlow Tables.

The rules for significant figures (pages xi to xv) are intended to be terse, direct, and simple, so that they may be easily acquired and retained. The strong type emphasizes the leading portions. The ordinary and finer types give details and explanations. For the sake of affording still greater prominence to the main working portions, some explanatory matter which will be unnecessary to many persons has been transferred from its more logical position of precedence to the latter part of the text. These rules in various forms have been in successful use by large classes of students, in connection with the author's "Physical Laboratory Notes" (printed, but not published, by the Massachusetts Institute of Technology), and his "Precision of Measurements." The recognition of the need of such rules amongst engineers and others whose practical work demands rapid and reliable computations was the cause of their general introduction into this laboratory instruction. It is therefore hoped that they may render effective service to others besides the students for whom they have been more directly designed.

In the arrangement of the tables, the effort has been exerted to make them correct, legible, systematic, and convenient in use. A new set of tables is, of course, liable to contain mistakes; notices of errata will therefore be thankfully received.

The special indexing of the corners of pages, the use of heavy type at points to be made conspicuous, the employment of spaces rather than rules for the partition of lines and columns, and the style of type and kind of paper used, are believed to conduce to legibility. As to system of arrangement, there are few novelties other than the insertion of the logs, cologs, and reciprocals of 1.000 to 1.100 at the top of the respective four-place tables, and the division of most of the four-place tables so that the second page begins with 5.0 instead of the customary 4.5. The frequent occurrence of correction and reduction factors, ranging from 1.0 to 1.1, renders this by far the most frequently used part of the table; while at the same time, on account of the large tabular differences, interpolation is here the most laborious. The insertion of logs, cologs, and reciprocals from 1.0 to 1.1 with increments of 0.001 and 0.0001, respectively, in the four- and five-place tables, obviates this interpolation. In tables of antilogs and square roots the addition would be of little service. In the tables of logarithms and of square roots, heavier type has been used at apparently scattered points throughout the body of the tables. These points, in the five-place logarithm tables, for instance, are where the first two figures in the mantissa change by one unit in the second place, *e.g.* 00, 01, 02, etc. The obvious service of this is to aid the eye in finding any desired mantissa in working the table backward to obtain the antilog or number corresponding. The object is, of course, the same in the other tables.

As to the wisdom of departing from the usual custom of omitting decimal points entirely from logarithm tables, the author believes that the retention of the point promotes clearness of comprehension of the tables by beginners, and lessens mental effort in more experienced computers, especially when associated with the notation by powers of 10, as in the explanations here given. It seems unfortunate that this simple notation, so useful in computation and so great an aid in the explanation of numerical relations, is not universally incorporated into arithmetical instruction.

The rules for the employment of logarithms and of the tables have not been prepared especially to meet the need of those entirely unfamiliar with the principles of logarithms, although they would probably be intelligible to any mature beginner. It is thought, however, that the explanations and instructions given may prove an aid even to those who are already somewhat familiar with the subject.

ROGERS LABORATORY OF PHYSICS,
MASSACHUSETTS INSTITUTE OF TECHNOLOGY.
Boston, August, 1896.

COMPUTATION RULES.

PROPER NUMBER OF PLACES OF SIGNIFICANT FIGURES.

The following three pages contain the rules and their underlying principles in a condensed form for ready reference. For readers to whom some of the terms employed are unfamiliar, or who desire fuller proofs and explanations, some additional pages of "Definitions and Explanations" have been appended.

These rules should enable a computer to decide at the outset of his work, or at the successive stages of it, what number of places of significant figures he should retain in order to avoid waste of labor on the one hand or sacrifice of accuracy on the other. They provide for a sufficient number of places to assure that (barring mistakes) the accumulated error arising from the rejection of further places shall be always smaller, usually much smaller, than the supposed uncertainty of the data or result, in computations involving not more than about 20 rejections. The retention of more places is worse than useless. It adds nothing to the accuracy of the result, although increasing materially the labor of computing, and the liability to mistake. The aggregate value of the time thus wasted, — obvious enough to any one who has had occasion to perform extended computations, — may be appreciated from the fact that the use of five places where four would suffice, nearly doubles the labor: using six places instead of four, nearly trebles it; thus wasting 100 and 200 per cent respectively of the necessary amount of work, and probably a greater proportion of time. Moreover, incongruities in the use of places of figures arouse skepticism as to the competence of the worker in other directions.

FUNDAMENTAL PRINCIPLES.

Retain everywhere enough places to correspond to two unreliable places in the final result; the direct object of this is to keep the first place of unreliable figures in the final result substantially free from the accumulated rejection errors.

EXCEPTIONS.—**A final result** is seldom stated to more than one uncertain place unless the uncertainty of that place is small (say plus or minus four or less).

Example: 1, page xvii.

Single direct measurements generally yield numbers extending to only one uncertain place. This should not, however, be taken as a reason for relaxing the application of the above rule to subsequent steps of the computation, especially in deducing the mean or average of several single observations.

Final zeros occurring in decimal fractions should be retained when any other digit in the same place would be retained. This is of course essential to show that this place is known.

The foregoing principles consistently carried out constitute entirely sufficient rules. But more detailed instructions are usually required at the outset. These are readily understood in view of the two following propositions, which one can easily verify by algebra or by numerical examples.

PROPOSITION I.—**In multiplication or division, the percentage accuracy of the product or quotient cannot exceed that of the factor whose percentage accuracy is least.**

PROPOSITION II.—**In addition or subtraction, the result cannot be accurate beyond the first decimal place which is inaccurate in any component.**

A more general form of statement from which these follow is: If several numbers are multiplied or divided, a given percentage error in any one of them will produce the same percentage error in the result. If several numbers are added or subtracted, a given error or change in the digit in any decimal or other place will produce an equal error or change in the digit in the same decimal place in the result.

RULES IN DETAIL.

REJECTION. — **In casting off places of figures, increase by 1 the last figure retained when the first left-hand rejected figure is 5 or greater; otherwise leave it unchanged.**

Example. — If the last two figures are rejected

	756 827.9	becomes	756 830.
and	0.00 263 439	becomes	0.00 263 4.

A MEAN OR AVERAGE **should always be carried to two unreliable figures.**

A mean is more reliable than the single observation from which it is computed (in proportion to $\sqrt{n}$, the square root of the number of observations). Thus, as the data frequently extend to only one unreliable figure, the mean will often have to be carried two places further than the single observation.

MULTIPLICATION OR DIVISION. — **Ascertain from the object of the work the percentage accuracy desired in the final result; or, by inspection of the data, find the percentage accuracy of that factor for which this is least,** *i.e.* for which the deviation-measure or the estimated error, expressed as a per cent, is largest. See Example 1, page xvii.

In direct multiplication or division retain in every factor, product, and quotient throughout the entire process, and in final results, for an accuracy of about

One per cent, or worse, four (4) places of significant figures;

One-tenth per cent, or worse, five (5) places of significant figures;

and so on.

In the ordinary and the shortened processes of "long multiplication," it is best to carry out the partial products one place beyond that yielding the last place required in the result under the above rule.

Examples: 2–5, page xvii.

LOGARITHMS. — **If the multiplication or division is performed by means of logarithms, the mantissa should contain as many places as are required by the foregoing rules for the direct process; i.e. for about**

One per cent, or worse, use four (4) place tables;

One-tenth per cent, or worse, use five (5) place tables.

Examples: 2–5, page xvii.

Addition or Subtraction. — **Ascertain from the stated object of the work the percentage accuracy desired. If this is about**

One per cent, or worse, carry the result to four (4) places of significant figures;

One-tenth per cent, or worse, carry the result to five (5) places of significant figures; and so on, and carry each component quantity to that place of decimals which would correspond to this required place in the result, that is, stand in the same column with that place.

Examples: 6–8, page xix.

When the desired accuracy is not stated, **inspect the data to find the component whose first uncertain place is furthest to the left, i.e. whose deviation measure** (page xlii), **in units, not percentage or fractional, is greatest. Retain this component to two uncertain places, and all other components to the place which would stand in the same column with this second place.**

Examples: 6–8, page xix.

N. B. — If the number of components approaches 20, care may well be taken in refined work that an unusually large rejection error does not enter through a special combination of rejected figures. The rules are, however, sufficient for the worst possible case.

The computer should notice that the *percentage* precision of a result which is the difference of two or more quantities will usually be smaller, and may be much smaller, than that of any of the component quantities.

Numerical Substitution in Formulæ. — A large number of formulæ may be represented by the type

$$x = \frac{a \cdot b \pm c \cdot d \pm \cdots}{p \cdot q \pm r \cdot s \pm \cdots},$$

where a, b, c, d, etc., represent numbers to be multiplied, divided, added, or subtracted, etc., as indicated. Any one or more of the factors and terms may be wanting; or, there may be several in place of two; and so on.

Obviously, in order that the result x shall be accurate to a specified per cent, both numerator and denominator must be at least of that accuracy, and each should therefore be carried out to the number of places of significant figures needed in x. Then as the numerator consists of two or more terms ab and bc added or subtracted, inspection under the foregoing rules for addition or subtraction will show to what decimal place each of these terms must be carried. Further, a and b must each be carried to the number of significant figures thus required in the product ab, and so on.

In complicated formulæ this process of inspection is sometimes slightly troublesome, but is essential unless the necessary precision of the components has been otherwise studied; as, for example, by the simple applications of the differential calculus as in the author's "Precision of Measurements."

Examples: 9–11, page xx. Practical examples of substitution in moderately simple formulæ.

NOTATION BY POWERS OF TEN.

Statement of the Method. — Regard the decimal point as merely an affix whose sole purpose is to indicate which is the units' place of figures. **Fix the attention firmly upon the units' place as the centre of symmetry of our customary system of notation.** The too universal reference to the decimal point, rather than to the units' place, in arithmetical rules and explanations, has resulted in masking this symmetry and in thus depriving the student of its important aid. In our common decimal system of notation, a digit in the units' place represents so many times unity, *i.e.* so many times $10^0 (=1)$, or so many *units*. In the first place to the left of the units' place the digit represents so many times 10^{+1}, *i.e.* so many *tens*, and in the first place to the right, so many times 10^{-1}, *i.e.* so many *tenths;* in the second place to the left so many times 10^{+2}, *i.e. hundreds*, and to the right so many times 10^{-2}, *i.e. hundredths;* in the sixth places, so many times 10^{+6} and 10^{-6}, *i.e. millions* and *millionths*, respectively: and so on. The fundamental symmetry of the whole system *about the units' place* is thus obvious, and should not be lost sight of.

In counting up places, whether to right or left, always begin with the units' place as zero.

It is clear, then, that we may write numbers in this way:

for 90 write $9 \cdot 10^1$;

for 6000 write $6.000 \cdot 10^3$ or $6 \cdot 10^3$ as the case may require;

for 345 write $3.45 \cdot 10^2$;

for 0.00 005 write $5 \cdot 10^{-5}$;

for 0.00 468 9 write $4.689 \cdot 10^{-3}$;

for 850.72 write $8.5072 \cdot 10^2$; and so on. That is,

Separate the number into two factors, the first being the original number with the decimal point changed in position so as to follow the first figure; the other being $10^{\pm n}$, where the sign is plus for a whole number and minus for a fraction, and where n is the number of places the decimal point has been moved.

To transform a number expressed in this way back into the ordinary form, move the decimal point n places, making the number a whole (or a larger) number if n is plus, and a fraction if n is minus.

Associate firmly in the mind the plus sign with whole numbers, the minus sign with fractions; thus avoiding confusion as to the sign of n.

In much work, the factoring need not be written out, but may merely be mental.

This notation reduces the error and work of locating the decimal point in multiplication or division, especially in expressions containing several terms in the numerator and denominator. It is very helpful in connection with the characteristic of logarithms, and the location of the decimal point in evolution, involution, and finding reciprocals. It saves space and promotes clearness in expressing large numbers or small fractions, and it is the best aid in following the decimal point while using the slide rule. It also enables one to dispense with characteristics in certain parts of computations (see Examples, page xxi).

An abbreviated notation helpful in one's own work, but perhaps not to be urged for general adoption, consists in dropping the $\cdot 10$, thus,

instead of $4.507 \cdot 10^2$ write merely 4.507^2
instead of $5.3704 \cdot 10^{-3}$ write merely 5.3704^{-3}

The adoption of the bracket or parenthesis, *e.g.* $(4.507)^2$, for either notation in cases of possible doubt removes all risk of mistaking these indices for ordinary exponents of powers.

Examples 9, 10, 11 give incidentally illustrations of the use of the notation by powers of 10.

Symmetrical Grouping of Figures. — For writing numbers, adopt the following system of groups and spaces: —

Write 143 258.64 796
instead of 143,258.647,96, the usual method.

A still clearer method would be to write

143 25**8**64 796

denoting the units' place by the heavy figure, but this is impracticable. The proposed system is symmetrical about the units' place, the customary system is not. It groups together the units, tens, and hundreds of thousandths, of millionths, etc., as well as of thousands, millions, etc. It is clearer and less liable to error by the substitution of spaces for the commas to mark off the groups Exception is usually to be made in the case of a decimal fraction containing only three or four places. Thus write 0.4612 rather than 0.46 12, and 6.382 rather than 6.38 2.

EXAMPLES.

Example 1. — Suppose that a final result was stated as 298549. ± 0.10 per cent; this would mean that its uncertainty or deviation-measure or estimated measure of accuracy (see page xlii) was ± 0.10 per cent. To how many places should it be retained? 0.1 per cent of the number is 0.001 × 300000 = 300. Therefore the last three places are uncertain, but as the uncertainty in the first left-hand one is small (3), two uncertain places should be retained. The result, therefore, should be written 298 550. ± 0.10 per cent.

Suppose a result given as 47.58 243 5 ± 0.0062. This would be an incorrect use of figures. The ± 0.0062 shows that the result is uncertain in the third and fourth, and therefore in all subsequent decimal places.* The fifth and sixth places of significant figures are thus unreliable, so that the seventh and eighth places are entirely valueless, and should, therefore, be rejected. We should be at liberty to use our judgment as to whether the result should then be written

$$47.58\,24 \pm 0.00\,62 \quad \text{or} \quad 47.582 \pm 0.006,$$

since the uncertainty in the fifth place is large. The second is more common practice. In this example the uncertainty is ± 0.0062/47. = ± 0.00 013, or ± 0.013 per cent. It might, therefore, have been expressed as ± 13 parts in 100,000, or ± 0.013 per cent instead of as ± 0.0062 units. It is always expressed in the same unit as the quantity itself, *e.g.* ft., lbs., etc., except when directly stated to be a percentage.

Example 2. — Desired with an accuracy of 2 per cent, the volume of a right circular cylinder whose radius r is 6.0428 inches, and length l 12.653 inches. Volume $V = \pi r^2 l$. By the rule for multiplication, since the result is desired to worse than one per cent, the data and all steps should be carried to four places of significant figures. Hence, we should have

$$V = 3.142 \times 6.043^2 \times 12.65 = 1451.$$

By ordinary multiplication:			By shortened multiplication:		
6.043	3.142	114.7	6.043	3.142	114.7
6.043	36.52	12.65	6.043	36.52	1265
18129	6284	5 735	36 258	94 26	1147
24172	1 5710	68 82	240	18 84	2294
36 258	18 852	229 4	18	1 55	690
36.52	94 26	1147	36.52	6	55
	114.7	1451.		114.7	1451.

* This quantity ± 0.0062, or whatever may be its value, is the "average deviation" or the "deviation-measure" of the result; that is, the average amount by which several results similarly obtained would differ from their mean. A fuller explanation is given at page xlii. The "probable error," which is nearly identical with the average deviation, is commonly used in its stead. Either suffices.

Observe that the partial products beyond the place standing over the fifth place of the result in each multiplication are useless. Hence the obvious saving of labor in the shortened process, which is also more compact. The process is easily understood by inspection of the example. Multiply first by the first left-hand figure of the multiplier. If the resulting partial product has one more place than is desired in the result, then drop the last figure of the multiplicand when multiplying by the second figure of the multiplier; drop the last two, when multiplying by the third figure; and so on. If, however, the first partial product has not the desired number, the dropping of figures must be deferred till the third figure of the multiplier is used.

Example 3. — Desired the volume $V = \pi r^2 l$ of a right circular cylinder whose dimensions are

$$r = \overset{\text{in.}}{6.0428} \pm \tfrac{1}{4} \text{ per cent}, \quad l = \overset{\text{in.}}{12.653} \pm \tfrac{1}{10} \text{ per cent}.$$

The result cannot be more accurate than the least precise factor, which is obviously r. Under the rule, $\frac{1}{4}$ per cent computations call for five places of significant figures. Hence we should have to find by multiplication or by five-place logarithms,

$$V = 3.1416 \times (6.0428)^2 \times 12.653.$$

Note in this connection that the error in V is proportional to twice the percentage error in r, since r enters in the second power, and that in general where a number is raised to any power n the percentage error in the result is increased to n times its value in the data. These rules, however, provide sufficiently for such cases. See "Definitions and Explanations."

Example 4. — Desired the volume $V = \pi r^2 l$ of a right circular cylinder whose dimensions are given as

$$r = 6.043 \pm 0.017 \text{ inches}, \quad l = 12.653 \pm 0.038 \text{ inches}.$$

Under the general principle of retaining places to correspond to two uncertain figures in the result in the data, r should have four places and l five places, judging from their stated precision. But the weaker quantity fixes the number of places in the result, so that we should use but four places:

$$V = 3.142 \times (6.043)^2 \times 12.65.$$

Example 5. — Desired the ratio of the radius to the length in each of the Examples 2, 3, and 4.

The number of places of figures to be used would be respectively four, five, and four, just as in the above solutions. A factor in the denominator follows precisely the same rule as to places of significant figures as the one in the numerator. To contrast the ordinary and shortened solutions, the following are given:

```
12.65) 6.043 (0.4777        |        6.043 | 12.65
       5060                 |        5060  | 0.4777
        9830                |        9830
        8855                |        8855
         9750               |         975
         8855               |         889
          8950              |          86
          8855              |          91
```

Example 6. — Desired the value to 0.6 per cent of

$$47.34\,89 + 174.32\,825 - 5.62\,147.$$

For 0.1 per cent or worse (page xiv), we must retain places in the components to correspond to five places in the result. By inspection we see that the result will be slightly more than 200. Hence its fifth place will be the second decimal place, and we need retain no place beyond that in the components. Thus,

$$\begin{array}{l} 47.35 \\ \underline{174.33} \\ 221.68 \\ \underline{-5.62} \\ 216.06 \end{array}$$

[± an unknown amount as precision-measure].

Example 7. — Desired the algebraic sum of

$47'.34\,89 \pm 0.0042$, $174'.32\,825 \pm 0.00\,089$, and $-5'.62\,147 \pm 0.00\,008$.

By inspection the weaker component, that is, the one whose first uncertain place is furthest to the left, is the first number. Retaining this to two uncertain figures would carry it to the fourth decimal place. It will then be useless to retain the other components beyond that place, and we shall have

$$\begin{array}{l} 47.34\,89 \\ 174.32\,83 \\ 221.67\,72 \\ \underline{-5.62\,15} \\ 216.05\,57 \end{array}$$

[± more than $0'.00\,42$].

Example 8. — Desired the algebraic sum of

$47'.34\,89 \pm 0.05$ per cent, $174'.32\,825 \pm 0.02$ per cent, $-5'.62\,147 \pm 0.1$ per cent.

0.05 per cent of 47. is 0.024; 0.02 per cent of 174. is 0.035; 0.1 per cent of 5.6 is 0.00 56. Hence the weakest component is now the second, and this, and consequently the others, should be retained to three decimal places. Thus, we have

$$\begin{array}{l} 47.349 \\ \underline{174.328} \\ 221.677 \\ \underline{-5.621} \\ 216.056 \end{array}$$

[± more than $0'.035$].

Example 9. — The horse-power, HP, which could safely be transmitted by a wrought-iron shaft of diameter d inches, running at a speed N rotations per minute, the safe shearing load of wrought iron being represented by f, is given by the expression

$$HP = \frac{2\pi^2 d^3 f N}{16 \cdot 12 \cdot 33000}.$$

[Deduced from Lanza's "Applied Mechanics," page 336. The several constants 2, 16, 12, and 33000 would of course be combined into a single constant in a working formula, but they are here left separate for purposes of better illustration.]

To how many places of significant figures should the quantities, result, and various steps of the computation be carried out to assure against a computation error in the result, sensible as compared to one per cent?

Solution.—In this and all similar problems, where the expression consists merely of a number of factors in the numerator and denominator (either or both), without additions or subtractions, the solution of the significant figure problem can be made without any knowledge of the magnitude of the component quantities, such as d, f, N, etc. In this example, as the result is desired to one per cent, according to the rules it should be carried to four places of significant figures. Hence, according to the rule, page xiii, or to Proposition II, page xii, each factor of the whole expression should be carried to four places. In this expression every quantity is a factor, either in the first or a higher power, viz. 2, π^2, d^3, f, N, 16, 12, and 33000. Each, therefore, should be carried to four figures. Hence, also, if direct multiplication be employed in the solution, each product and quotient must be carried to four places. If logarithms are used (they should be) four-place tables should be chosen. When a quantity enters as a factor of the nth power this is equivalent to its entering n times as a simple factor or as n separate factors, all with the same percentage error of the same sign. See also note under example 3.

The constants 2, 16, 12, and 33000 do not require to be carried to more places than they are here given because they are complete as they stand, that is, all further figures are known to be zero as a matter of definition or mathematical fact. If either of them had been an experimental constant, that is, determined by measurement, it should have been carried out to four places even if the last figure or two were zero. For instance, if experimental, the 16 should have been written 16.0, 16.00, 16.000, and so on according to the number of places to which it was known (see rule, page xii). Failure on the part of those who write such formulæ to adhere to this convention, or to indicate in some clear way the degree of accuracy possessed by such constants, is a serious source of annoyance and trouble to those who use them.

As elsewhere it must not be inferred if certain of the quantities, *e.g.* d or f, in this expression cannot be carried out to this desired number of figures, that consequently the result will not have the accuracy desired in the given case. The outcome of such a condition would merely be that these factors *would be liable* to introduce more than a safe computation error. For instance, if f were given as 10100 lbs. per square inch, we should have no certainty that it was carried far enough. The presumption would be that it was correct to but three places, and therefore not exact enough. If, however, from a knowledge of the subject we were aware that the best known value was 10110, we should know that the error from using 10100 was only 1 in 1000 or 0.1 per cent, and hence admissible. On the other hand, if we know that the best value was 10050, we should know that the computation error in the result from using 10100 was 0.5 per cent, and hence by no means safe in the above problem.

More complete methods for ascertaining the exact accuracy needed in each component measured quantity in such formulæ, are given in the author's "Precision of Measurements." It is to be remembered that we are now dealing merely with rules for *computation* errors, and these are not suited to the other problem. They are intended to secure a *safe number of places for the worst case*, and would, therefore, impose unduly severe requirements as to the accuracy necessary in the *measurement* of the components in most cases.

Numerical Substitution. — The numerical expression to be solved if written out would be

$$\frac{2 \cdot 3.142^2 \cdot 1.364^3 \cdot 10\,000 \cdot 300}{16 \cdot 12 \cdot 33\,000}$$ in the ordinary notation;

$$\frac{2 \cdot 3.142^2 \cdot 1.364^3 \cdot 10^4 \cdot 3 \cdot 10^2}{1.6 \cdot 10^1 \cdot 1.2 \cdot 10^1 \cdot 3.3 \cdot 10^4}$$ in the notation by powers of 10 (page xv);

$$\frac{2(3.142)^2(1.364)^3 10^4 \cdot 3^2}{1.6^1 \cdot 1.2^1 \cdot 3.3^4}$$ in the abbreviated notation by powers of 10 (page xv).

The first would be worked out in the usual manner by direct multiplication or by logarithms, as shown below.

The second would be worked as follows:

Multiply together the terms other than 10^n of the numerator, *i.e.*	$2 \times 3.142^2 \times 1.364^3 \times 3 = 150.3$
Multiply together the terms other than 10^n of the denominator, *i.e.*	$1.6 \times 1.2 \times 3.3 = 6.336$
Divide numerator by denominator	$= 23.72$
Add together all indices of powers of 10 in numerator, also in denominator, and subtract the latter from the former. Or, better, add (algebraically) all the indices, reversing the sign of those in the denominator, thus: $4 + 2 - 1 - 1 - 4 = 0$.	
The result is therefore $23.72 \cdot 10^0$	$= 23.72$

Note distinctly that all this writing out of the fraction and of the several steps is merely for the purpose of this explanation. In an actual solution such of these operations as are essential to the work would be conducted mentally, the actual multiplication and division alone being written out.

If the solution were made by logarithms, it might assume either of the two following forms. The first is the usual one, the second shows how the use of the powers of 10 enables us, if we choose, to dispense with writing characteristics in very many places, — a saving of just so much labor. The factor 10^0 in the second result is, of course, obtained by summing mentally the indices of the factors 10, those in the denominator being taken with reversed sign, as in the preceding paragraph. These indices would not be written out, but taken by direct inspection of the data as originally written.

Denominator.	Usual Method.	Dropping Characteristics.
log 16.	= 1.2041	.2041
log 12.	= 1.0792	.0792
log 33 000.	= 4.5185	.5185
log denom.	= 6.8018	.8018

Numerator.	Usual Method.	Dropping Characteristics.
log 2.	= 0.3010	.3010
2 × log 3.142	= 0.9944	.9944
3 × log 1.364	= 0.4044	.4044
log 10 000.	= 4.	
log 300.	= 2.4771	.4771
	8.1769	2.1769
	6.8018	.8018
	1.3751	1.3751
Result,	23.72	$23.72 \cdot 10^0$

Example 10.—The crushing load of a hollow, cast-iron pillar of circular section, with concentric surfaces of diameters D and d as given by Hodgkinson (Lanza, "Applied Mechanics," page 332) is

$$c = 109\,801 \frac{\pi(D^2 - d^2)}{4}.$$

Desired to ten per cent the load which a pillar of external and internal diameters 4.032 inches and 2.16 inches, respectively, would carry. How many places of figures should be used in the computation?

Ten per cent results call for three figures in all factors (page xiii). The factors in this expression are 109 801, π, $(D^2 - d^2)$, and 4, each of which should therefore be carried to three figures. The first two should therefore be 110 000 and 3.14. The 4 is a complete number as it stands. $D^2 - d^2 = 4.0^2 - 2.2^2$ roughly $= 16.0 - 4.8 = 11.2$. To have three figures, it should therefore be carried to the first decimal place. Then as it is made up of two quantities, one subtracted from the other, each of these should be carried (page xiv) to the decimal place desired in the result, *i.e.* to the tenths' place. This requires D^2 to contain three figures, 16.0, and hence D *should* contain three figures (since D^2 consists of the factors $D \times D$), *i.e.* should be written 4.03 inches. The requirement of one decimal place in d^2 calls for but two figures, 4.8, and hence two figures, 2.2 inches, in d. The numerical expression to be solved would then be

$$c = 110\,000 \frac{3.14(4.03^2 - 2.2^2)}{4},$$

which would be most easily worked by a simple slide rule, or by direct multiplication.

Example 11.—Desired to 0.1 per cent the fraction of dry steam in a sample of steam, using the following observations made with the "Barrus Calorimeter" (Peabody's "Themodynamics of the Steam Engine," page 234), the formula being

$$x = \frac{W(q_2 - q_1) + e - w(q - q_3)}{wr},$$

where

x = fraction of a mixture which is dry steam	
W = weight of cooling water	573 5 lbs.
w = weight of condensed water	29.89 lbs.
t = temperature of steam	
t_1 = initial temperature of cooling water	37°.49 F.
t_2 = final temperature of cooling water	83°.84 F.
t_3 = temperature of condensed water	304°.88 F.
q_3 = "heat of liquid" at $t°_3$, from t_3 and tables	274.4
q_2 = "heat of liquid" at $t°_2$, from t_2 and tables	51.91 B. T. U.
q_1 = "heat of liquid" at $t°_1$, from t_1 and tables	5.53 B. T. U.
q = "heat of liquid" at $t°$, from t and tables	287.6 B. T. U.
e = radiation loss during test	120. B. T. U.
r = latent heat of steam at $t°$, from tables	891.2 B. T. U.

What number of places of figures should be used throughout?

Solution.—For 0.1 per cent the result, and therefore all factors, should have five places of figures (page xiii). The only factors of the whole expression are the whole numerator, w, and r. They should therefore be carried to five places. Note, however, that w and r are not so given in the data. Whether, however, they are given closely enough must be determined by other means (see remark at foot of page xx). Their product must, however, be carried to five places, and five-place log tables should be employed. The numerator consists of three terms, whose values, roughly, are

$$570(52. - 6.) = 2600.,\ 120.,\ \text{and}\ 30(288. - 274.) = 420.$$
$$\therefore\ \text{numerator} = 2600. + 120. - 420. = 2300.,\ \text{roughly}.$$

To have five places the numerator must be carried to one place of decimals, as also must each of its terms. The first term is composed of two factors, W and $(q_2 - q_1)$, each of which must therefore be carried to five places. Then as $q_2 - q_1 = 52. - 6 = 46.$ roughly, it must be carried to the third decimal place. Hence q_2 and q_1 must each be carried to the third decimal place. The third term, in order to extend to the first decimal place, must contain four figures. It consists of two factors, w and $(q - q_3)$, each of which must thus contain four figures. To make the value of $q - q_3$ contain four figures, its numerical value, 14, must be carried to the second decimal place. This would require that both q and q_3 be carried to the second decimal place, or to five figures each.

To summarize, then, putting in an interrogation point wherever a figure is wanting in the data, we have as the numerical expression to be solved

$$x = \frac{573.5?\,(51.91? - 5.53?) + 120.? - 29.89(287.6? - 274.4?)}{29.89? \times 891.2?} = 0.988.$$

Obviously, then, on inspection, although we might carry out the products $W(q_2 - q_1)$, $w(q - q_3)$, and wr, each to the necessary five figures, much doubt is cast upon the sufficiency of the data themselves to give the desired 0.1 per cent in the result. The problem would require a detailed study by other methods to decide that point, — the result of which, it may be incidentally asserted, would be adverse.

Note. — It may not be amiss in connection with these problems to call attention to the very large amount of engineering computations in which four, often three (slide rule), places of figures are abundant. In the design of machines and structures, the strength and sizes of parts, such as struts and tie rods, beams, pillars, shafting, etc., and the strains or stresses in them, often cannot, or need not, be fixed upon within an accuracy of one or even many per cent. This limit is fixed by the unreliability of materials or workmanship, or by ignorance of the exact conditions to which the parts may be subjected. Like uncertainties affect many of the data upon which specifications, estimates, and contracts for engineering work are based, and the experimental constants in sundry formulæ. More than three or four places of figures can be indulged in for such work only at an extravagant waste of time. On the other hand it is necessary to discriminate sharply such operations as linear, angular, surface, and sometimes levelling, measurements in surveying and geodetic work where the accuracy may be very high.

LOGARITHMS.

Before reading the following pages become familiar with the "Notation by Powers of 10," page xv.

The **common or Briggs logarithm** of a given number is the exponent of that power to which 10 must be raised to produce the number. Thus, 3 is the common logarithm of 1000, since $10^3 = 1000$.

To multiply numbers together, add their logarithms. The sum is the logarithm of the desired product.

Proof.—The product $10^a \times 10^b \times \cdots \times 10^m$ is $10^{a+b\cdots+m}$, since powers of the same number may be multiplied together by adding their exponents. Therefore, if $A = 10^a$, $B = 10^b$, $\cdots$, $M = 10^m$, $A \times B \times \cdots \times M = 10^{a+b+\cdots+m}$. That is, $a + b + \cdots + m$ is the logarithm of $A \times B \times \cdots \times M$. But a, b, $\cdots$, m are respectively $\log A$, $\log B$, $\cdots$, $\log M$. Hence, $\log (A \times B \times \cdots \times M) = \log A + \log B + \cdots + \log M$.

To divide one number by another, subtract the logarithm of the latter from that of the former. The difference is the logarithm of the quotient.

Proof. — Following the foregoing notation, $A/B = 10^a/10^b = 10^a \times 10^b = 10^{a-b}$. Hence $\log A/B = a - b = \log A - \log B$.

Logarithm of a number of several figures. — As $1 = 10^0$ and $10 = 10^1$, the logarithm of 1 is 0 and of 10 is 1, and the logarithm of any number greater than 0 and less than 10, that is, of any number in the units' place (whether or not followed by a decimal fraction) is less than 1, that is, it is a fraction. It is expressed as a decimal fraction.

From the definition of a logarithm it is obvious that the logarithm of any stated power of 10 is the index of the power; *i.e.* $\log 10^{\pm n} = \pm n$ when $\pm n$ is any number, whole or fractional, positive or negative. Hence, taking first a specific example, the logarithm of 306.2, being of course the same as the logarithm of its equal, is the same as the logarithm of $3.062 \cdot 10^2$ (see "Notation by Powers of 10," page xv), which is, $\log 3.062 + \log 10^2 = .4860 + 2.$, which is usually written 2.4860. The .4860 is found from tables, as shown later.

Similarly, $\log 0.00\,306\,2 = \log 3.062 \cdot 10^{-3} = \log 3.062 + \log 10^{-3} = .4860 - 3.$, which is usually written either $\bar{3}.4860$ or $7.4860 - 10$, as will be further explained.

The logarithm then consists of, or may be separated into, two parts, viz. first, the decimal part called the **mantissa**, which is the logarithm of the first factor in the above separation; second, the integral part, or whole number, preceding the decimal point, and called the **characteristic or index**, which is the logarithm of the second factor $10^{\pm n}$, and which, therefore, is $\pm n$.

Tables of logarithms contain the logarithms of the numbers from 1. to 10., by steps larger or smaller, and to as many decimal places as may be requisite for the accuracy sought in the work in which they are to be employed. But all numbers whatever, from 0 to ∞, are one of these numbers 1. to 10. multiplied by some power of ten, *i.e.* by $10^{\pm n}$. For example,

$$4\,628\,326 = 4.62\,832\,6 \cdot 10^6, \text{ and } 0.03\,986 = 3.986 \cdot 10^{-2}.$$

Hence, the tables enable one, by merely prefixing to the tabular value the proper "characteristic" $\pm n$, to obtain the logarithm of any number whatever, from zero to infinity. The quantity directly given in the table is obviously the mantissa of the desired logarithm, and is therefore always a decimal fraction.

Since any table gives the mantissa to only a specified number of decimal places, it can represent only a correspondingly restricted number of places of significant figures in the original number. It is to be remembered that a change of one unit in the last decimal place of the mantissa corresponds at all points throughout a table to a constant percentage change ("Definitions and Explanations") in the number corresponding. The amount of this change is such that it becomes the proper custom to use logarithm tables giving the mantissa to a number of places equal to the number of significant figures retained in the original quantities. Thus, if the numbers entering into the computation are properly retained to four places of significant figures, a four-place logarithm table should be used in connection with them; if to five significant figures, a five-place table; and so on.

Four-place logarithm tables contain the logarithms to four decimal places of all numbers of three figures from 1.00 to 9.99, and enable one by interpolation to obtain the four-place logarithm of any four-place number from 1.000 to 9.999. By merely prefixing the proper characteristic $\pm n$, therefore, the four-place logarithm of any four-place number from 0 to ∞ is obtained, or, in other words, the four-place logarithm of any number whatever from 0 to ∞ in so far as this is governed by the first four significant figures of the number. Four-place tables should not be employed upon work of an accuracy exceeding one-half of one per cent.

Five-place tables give directly the logarithms to one place further than four-place tables, *i.e.* to five decimal places, for numbers from 1.000 to 9.999, and thence, by interpolation, from 1.0000 to 9.9999. Thus, with the proper characteristic, these tables furnish the logarithms of all five-figure numbers from 0 to ∞, that is, of any number whatever in so far as this is governed by the first five significant figures of the number. Five-place tables should not be employed in work of an accuracy exceeding one-twentieth of one per cent.

Six-place tables are sometimes arranged with the same steps as five-place, *i.e.* giving directly the logarithm to six decimal places of numbers from 1.000 to 9.999 only. Such tables are of no practical service; for it is entirely useless to employ six-place logarithms in work on five-place numbers, and interpolation for six-place numbers in tables of so large steps as this, besides being less reliable, is more laborious and annoying than is the use of the more bulky tables of smaller steps.

If six-place tables are desired, it is usual to employ, dropping the last place, tables which give directly seven-place logarithms of numbers from 1.0000 to 9.9999 with convenient interpolation tables for the next place. Of these, the Vega tables are among the most convenient, legible, and reliable, being also comparatively inexpensive. The seventh place is very rarely demanded by physical, chemical, or engineering work.

The relative labor in using four, five, and six place tables lies probably between the ratios 1 : 2 : 3 and 2 : 3 : 4. Assuming the first estimate to hold, the labor is doubled by using a five-place instead of a four-place table, and is increased one-half by using a six instead of a five place table. Hence, as there is no sensible gain from using an excess of places, it is obviously very important to employ a table of the smallest admissible number of places. But, on the other hand, the use of too few places must be guarded against. As an instance of a somewhat dangerous practice may be cited the use of four-place

tables in 0.1 per cent work. This is a not infrequent practice, most common perhaps in chemical computation, and, of course, arising from the exceeding convenience of cards containing four-place logarithms. It will be shown at page xliv, that four places are not sufficient for 0.1 per cent direct computations, and the error if four-place logarithms are used is sensibly the same. The computation error may easily rise to 0.2 or 0.3 per cent with four-place tables even in ordinary computations.

To Find the Logarithms of a Number.

RULE. — **Regard the number Q as separated into two factors $q \times 10^{\pm n}$, where q begins in the units' place (see "Notation by Powers of 10," page xv). Find in the tables the logarithm of q. This will be the mantissa of the desired logarithm. Prefix to this the characteristic or index $\pm n$.**

A few examples will sufficiently elucidate the process.

Example. — Desired the logarithm to four decimal places of the number 306.

Write the number, or, better, merely consider it as if factored in the form $3.06 \cdot 10^2$. In the four-place table, on the line 3.0 and in the column headed 6 will be found .4857, which is log 3.06. Obviously, $\log 10^2 = 2$. Therefore $\log 306. = \log 3.06 + \log 10^2 = .4857 + 2$, which is usually written 2.4857.

Further Examples. Interpolation. — Desired to four decimal places the logarithm of 306.2.

This will lie between the logs of 306 and 307 (and approximately 0.2 of the way), and as the table is not carried out further we must interpolate.

For 3.07 we find .4871	Difference = .0014, usually written 14.
For 3.06 we find .4857	The desired mantissa will be 0.2 of the way from .4857 to .4871. Hence we must add to the former number $0.2 \cdot 14. = 3$.
Interpolation, 0.2 of 14 = 2.8 = 3	
$\therefore \log 3.062 = .4860$	
$\therefore \log 306.2 = \log 3.062 + \log 10^2 = 2.4860$	

The interpolation may always be made by subtracting and multiplying as in this example, but to save the labor, logarithm tables are usually provided with marginal **interpolation tables**, by the aid of which interpolations may easily be made mentally.

Thus in taking out log 3.062 we find log 3.06 = .4857. By inspection, difference = 14. In the interpolation table headed 14, line 2, stands 3, which is therefore the desired 0.2 of 14. Therefore log 3.062 = .4860. This operation could, of course, be carried out mentally.

The present tables are arranged so that the interpolation tables stand opposite, or nearly so, to the logarithms to which they correspond. This not only gives them a convenient location but enables the computer usually to avoid even the mental subtraction of the successive logarithms to find the difference, since this will, of course, be that at the head of the nearest difference table. Usually also the error introduced by using an interpolation table slightly too large or too small is negligible.

Interpolation becomes more laborious and more liable to error, if conducted mentally, in proportion as the difference is large. It is therefore greatest in the first quarter of any logarithm table. But it happens that in physical, chemical, and engineering computations there very often enter correction or reduction factors and other terms of the form $(1 + a)$ or $1/(1 - a)$, where a is a decimal fraction rarely as large as 0.1. The frequency of such terms calls for a disproportionately large number of the more laborious interpolations. To avoid this labor and increased chance of error, the excellent practice has been adopted in some five-place tables of inserting two pages giving the logarithms from 1.0 to 1.1, by steps only one-tenth as large as in the rest of the tables, thus doing away with all interpolation in this most used and most troublesome portion of the table, without adding seriously to its bulk. Such a table has here been prefixed to the five-place table. In the four-place table the same result has been accomplished here, in a manner which is perhaps novel, by the insertion of ten additional lines at the head of the table. The Vega seven-place tables unfortunately lack this feature.

To find Logarithm of a Decimal Fraction. — The procedure is precisely the same as for a whole number. Note that the logarithm of a decimal fraction is always negative, and, conversely, that a negative characteristic always denotes a decimal fraction.

Example. — Desired the log of 0.00 306 2.

$$\begin{aligned} \log 0.00\,306\,2 &= \log 3.062 \cdot 10^{-3} \\ \log \quad 3.07 &= .4871 \\ 3.06 &= .4857 \quad \text{Difference} = 14. \\ 0.2 \text{ of difference} &= \quad 3 \\ \therefore \log \quad 3.062 &= .4860 \\ \log \quad 10^{-3} &= -3. \\ \therefore \log 0.00\,306\,2 &= .4860 - 3. \end{aligned}$$

This is written either $\bar{3}.4860$ or $7.4860 - 10$. The latter form is obtained by adding 10 to the characteristic and appending -10 to the whole. The numbers thus appended must in many instances, but not in all, be followed up in the computation in order to correctly locate the decimal point at the close. This is, however, usually very little trouble. The second method is the very general practice and is based on the assertion that when several logarithms are to be added, it is more convenient to have all the characteristics positive. The author is, however, of the opinion that this conventional method serves almost no useful purpose, and that it is better and less troublesome in every way to retain the negative characteristic. It is almost as easy to add a column of numbers in which some are negative, and are therefore subtracted when they are reached instead of being added, as to add a column in which all are positive. And if the negative characteristic is retained, all care and writing of − 10 or its multiples is avoided. Moreover, the logarithm is complete in itself and shows at once that it is the log of a decimal fraction. These remarks apply not only to the logarithms of ordinary numbers but to the logarithms of the trigonometric functions.

Example of addition where some of the characteristics are negative. Places beyond the first of the mantissa are not added in this illustration.

2.4036	Beginning at the bottom of the columns we have
$\bar{1}.2168$	$5 + 3 = \mathbf{8} + 2 = \mathbf{10} + 4 = \mathbf{14}$, write 4;
1.3462	carry $\mathbf{1} + \bar{2} = \bar{\mathbf{1}} + 1 = \mathbf{0} + \bar{1} = \bar{\mathbf{1}} + 2 = \mathbf{1}$.
$\bar{2}.5113$	Of course the figures printed in heavy type are the only ones pro-
1.4	nounced or mentally enforced.

Examples of some cases where the use of the notation by powers of 10 enable one to dispense with the characteristic in portions of computations are given incidentally at page xxi.

Grouping by Fives in the Tables. — Throughout the entire set of tables, it will be observed, the columns and lines are arranged in groups of five. This not only aids the eye to readily follow any desired line or column, but enables the computer with a little practice to enter a desired column or line without glancing at the number at the top of the column or side of the line, and similarly to read off the number of column or line without glancing off to the marginal figure. Thus the middle column in the second group is 7, the last in the first group is 4, and so on. The practice of working by observed position rather than by the marginal number should be pursued. It reduces fatigue and tends to prevent mistakes.

Indexing at Corners of Tables. — The bold type at the outer corners of the tables shows the contents of the page and its opposite. Thus on the sixth page of the five-place log table, the larger black figure **2** shows that these two pages contain the logs of all the numbers beginning with 2. The smaller black figures, **30, 47,** in the margin show that the mantissæ on the two pages extend from 30 to 47 (first two figures). The indexing will be found a material help in rapidly running the page corners under the finger to find either a desired table or a desired figure in a table.

Decimal Point in Logarithm Tables. — It is the almost universal custom to omit the decimal point entirely from logarithm tables. This tends toward compactness, but is by no means essential on that score. The omission causes no serious inconvenience after slight practice. On the other hand, the retention of the point renders the table complete and strictly self-consistent; that is, any mantissa in the table is then the completely expressed logarithm of the corresponding tabular number. This fact tends materially toward clear comprehension on the part of the beginner or occasional computer. The table is also then perfectly assimilated to the "Notation by Powers of 10," page xv, which affords not only the clearest basis of explanation of mantissa and characteristic, but by far the easiest method of obtaining and following the characteristic in computations. The points do not add necessarily to the bulk of the tables and are no hindrance to their use by computers accustomed to other rules than those here given respecting the characteristic. They are therefore retained in these tables.

Antilogarithm = Number Corresponding. — Given the logarithm to find the "number corresponding," *i.e.* the number of which this is the logarithm. This number is also called the "antilogarithm."

From the Log Tables. —

Example. — To find the antilog of 2.4857, inspect the body of the four-place log table to find the mantissa .4857 (or the next smaller than this if the exact value does not appear). It will be found to lie on line 3.0 and in column 6, and is, hence, the log of 3.06. The characteristic 2 is the log of 10^2.

$$\therefore \text{antilog } 2.4857 = 3.06 \cdot 10^2 \text{ or } 306.$$

Example. — To find the antilog of 2.4860. In the table this mantissa does not appear exactly, but the next smaller is .4857, whose antilog is 3.06, and the difference from this to the next larger is 14.

$$.4860 - .4857 = 3, \text{ and } \tfrac{3}{14} = .2,$$

$$\therefore \text{antilog } .4860 \text{ is } \tfrac{3}{14} \text{ or } 0.2 \text{ of the way from } 3.06 \text{ to } 3.07,$$

$$\therefore \text{antilog } .4860 = 3.062,$$

also $$\text{antilog } 2. = 10^2,$$

$$\therefore \text{antilog } 2.4860 = 3.062 \cdot 10^2 \text{ or } 306.2.$$

The interpolation can be mentally made by the marginal interpolation tables. The tabular difference is 14, and it is desired to know what decimal fraction the difference 3 is of this. Looking down the column under 14 for the number nearest to 3 it is found to be 3 and to stand opposite to 2. ∴ 3 is 0.2 of 14, and antilog .4860 = 3.062.

It will be noticed that in the four-place and five-place log tables those mantissæ have been printed in heavier type in which the first figure changes from one digit to the next. This serves as a guide to the eye in looking for any desired mantissa whose antilog is sought.

From Tables of Antilogarithms. — Some computers prefer to employ a special table for antilogarithms instead of working backward in the ordinary logarithm table as in the preceding example. Whether the small saving in time effected by this means is an equivalent to the slight additional mental effort incident to the employment in alternation of tables differently arranged, may be questioned. Where many antilogs are to be taken out in succession, the gain is, however, sensible.

Example. — To find by the table of Antilogarithms the antilog 2.4857.

Taking the mantissa first, we find on line .48, column 5, the next smaller number 3.055 with a tabular difference of 7. Hence, antilog .485 = 3.055. But antilog .4857 will be 0.7 of the way from that of .485 to .486. The amount to be added then for interpolation is $0.7 \times 7 = 5$.

$$\begin{aligned} \text{antilog } .4857 &= 3.055 + 0.005 = 3.060, \\ \text{antilog } 2.4857 &= 3.060 \cdot 10^2 \quad = 306.0. \end{aligned}$$

Interpolation tables are provided whose method of use is sufficiently obvious.

Cologarithms = Logarithms of Reciprocals. — The colog of any number p is the logarithm of the reciprocal of the number.

It is therefore $\log 1 - \log p = 0 - \log p$.

In substitution in a formula such as $x = \frac{a \cdot b \cdots}{c \cdot d \cdot e \cdots}$, which is the product of several numbers (one or more) divided by the product of several others (one or more), the direct process would be to take the sum of the logarithms of the factors in the numerator, and to subtract from this the sum of the logs of the factors in the denominator.

Example. —

$$\begin{aligned} \log a &= 1.2543 & \log c &= 3.8642 \\ \log b &= \bar{3}.8766 & \log d &= \bar{5}.2121 \\ & & \log e &= 2.1345 \\ \text{Numerator sum} &= \bar{1}.1309 & \text{Denominator sum} &= 1.2108 \\ \text{Subtract denominator sum} &= 1.2108 \\ \text{Difference} &= \bar{3}.9201 \end{aligned}$$

x = number corresponding = $8.320 \cdot 10^{-3}$, or 0.00 8320.

This process may be simplified by employing the colog. It then becomes

$$\begin{aligned} \log a &= 1.2543 \\ \log b &= \bar{3}.8766 \\ \text{colog } c &= \bar{4}.1358 \\ \text{colog } d &= 4.7879 \\ \text{colog } e &= \bar{3}.8655 \\ \text{Sum} &= \bar{3}.9201 \end{aligned}$$

x = number corresponding = $8.320 \cdot 10^{-3}$

This process is thus reduced to the simple addition of a series of numbers.

The colog may be easily taken out of the usual log table by merely looking out the logarithm and subtracting mentally from 0.

Example. — Desired the colog of 306.2. Log 306.2 = 2.4860, colog = $\bar{3}$.5140.

The use of cologs either with or without a table effects but small saving, except in case a series of substitutions in a given formula are to be made, so that a number of cologs may be looked out in immediate succession.

Table of Cologs. — The table of four-place cologarithms is arranged similarly in every respect to the table of four-place logarithms, and the cologs are taken from it just as logs from their table; noting, however, that the cologs in the table have the characteristic $\bar{1}$, and that the differences are subtractive.

Example. — Desired the colog of 306.2. Separate into $3.062 \cdot 10^2$.

In line 3.0, column 6, colog 3.06	= $\bar{1}$.5143	Difference from
0.2 × (− 14) by difference table	= −3	colog 3.07 is − 14.
∴ colog 3.062	= $\bar{1}$.5140	
colog 10^2	= $\bar{2}$.	
∴ colog 306.2 = colog 3.062 + colog 10^2	= $\bar{3}$.5140	

Example. — Desired the colog of 0.00 713 6.

In line 7.1, column 3, is colog 7.13	= $\bar{1}$.1469	Difference is − 6.
0.6 (− 6) by difference table	= −4	
colog 7.136	= $\bar{1}$.1465	
colog 10^{-3}	= + 3.	
colog 0.00 713 6 = colog 7.136 + colog 10^{-3}	= 2.1465	

Habit in Reading off Numbers or Logarithms. — Time can be economized, strain on the attention reduced, and liability to mistake lessened by an easily acquired habit of grouping and emphasizing the figures in reading off the numbers, the mantissa, and the numbers corresponding (antilogs) in using tables.

A good method of reading is as follows:

In reading a number or antilog pay no regard to the decimal point. Emphasize the first figure; pause, read second and third figures; pause, read remaining figures in groups of three. Thus: Desired the log of 30.62 047 2. Read this as 3 06 204 72, *i.e.* three . . . naught six . . . two naught four . . . seven two.

In reading off a mantissa use no emphasis, but group the first two figures together, and the subsequent figures by threes. Thus, the mantissa .4869 would be read as 48 69, *i.e.* forty-eight . . . sixty-nine; and .48601 as 48 601, *i.e.* forty-eight . . . six naught one.

In taking from the log table the numbers corresponding (antilog) to .48601, it would be read 3 06 2, three . . . naught six . . . two, precisely as under the above rule for reading a number.

These rules apply equally well to four or five place tables. In tables of six or seven places the first three figures of the mantissa are grouped together in printing, and are therefore more conveniently read off together. Also, it is more convenient to read off six-place numbers and antilogs with three figures instead of two in the second group; thus, 7 814 62.

The difference between a number and a mantissa thus read off, whether audibly or mentally, almost precludes the possibility of mistaking one for the other, so that less strict attention will be required to avoid entering the number column of a table with a mantissa, or *vice versa*. It also avoids the mental or verbal employment of words of instruction. Thus, if a computer reads off 7 82 he knows, or his assistant using the log tables knows, as soon as the first figure has been read that the logarithm of the number is desired. Conversely, if he reads off 43 857, the first two figures alone show that the quantity is a mantissa, and that the antilog is required. Thus, no words of instruction need be used throughout an entire computation, and yet no possibility of error need enter.

It is also to be noted that this grouping is consistent with the symmetrical grouping advocated at an earlier page; adapts itself perfectly to the employment of the notation by powers of 10; and coincides with the most convenient grouping in the five-place tables.

Powers and Roots by Logarithms. — If $a = 10^m$ (so that $m = \log a$) then $(a)^n = (10^m)^n$ and $\log(a^n) = n \log a$. Here both m and n may be either positive or negative, and either an integer or a fraction.

Hence, to raise any number, whole or fractional, to any power, integral or fractional, and positive or negative, multiply the logarithm of the number by the exponent of the power of the number.

Since a root is a fractional power, *i.e.* $\sqrt[n]{a} = a^{\frac{1}{n}}$, the above rule includes the extraction of a root. In the case of decimal fractions the fact that the characteristic is negative while the mantissa is positive, must be regarded. The correct result will be assured if the mantissa and characteristic are separately multiplied (or divided) by the exponent, and the latter result then subtracted from the former, as will be further shown. Certain special procedures will also be described.

The one exception to this is where the index of the power is a single figure and positive. This case takes care of itself without special attention.

It may seem that the use of the negative characteristic thus increases labor over the usual notation explained at the foot of page xxviii. Inspection of examples will show that the only case in which it does so is where the index of the power or root is other than a single digit.

Powers and Roots of Numbers greater than Unity.—

Examples.—Desired the cube of 47.16, *i.e.* $(47.16)^3$.

$$\begin{array}{ll} \log 47.16 & = 1.6735 \\ \text{Multiply by} & \quad\underline{\quad 3} \\ \log x & = 5.0205 \quad \therefore x = 1.048 \cdot 10^5, \text{ or } 104\,800. \end{array}$$

Desired $\sqrt[5]{471.6}$.

$$\begin{array}{ll} \log 471.6 & = 2.6735 \\ \text{Divide by} & \quad\underline{\quad 5} \\ \log x & = 0.5347 \quad \therefore x = 3.425. \end{array}$$

Example.—Desired the 2.416 power of 65 830.

$$\begin{array}{llr} \log & 65\,830. = & 4.8184 \\ \text{Multiply by} & & \underline{2.416} \\ & & 9\,6368 \\ & & 1\,92736 \\ & & 4818 \\ & & \underline{2892} \\ \therefore \log & (65\,830)^{2.416} = & 11.6413 \\ \therefore & (65\,830)^{2.416} = & 4.378 \cdot 10^{11} = 437\,800\,000\,000. \end{array}$$

To avoid the direct multiplication of the logarithm by the index of power, when this contains several figures, their logs may be taken. Thus

$$\begin{array}{rlll} & & \text{(5 pl. logs.)} & \text{(4 pl. logs.)} \\ \log\log 65\,830 = \log 4.8184 & = & 0.68\,291 & 0.6829 \\ \log 2.416 & = & \underline{0.38\,310} & \underline{0.3831} \\ \log\log (65\,830)^{2.416} & = & 1.06\,601 & 1.0660 \\ \text{antilog } 1.06\,601 = \log (65\,830)^{2.416} & = & 11.641 & 11.64 \\ \therefore (65\,830)^{2.416} & = & 4.376 \cdot 10^{11} & 4.365 \cdot 10^{11}. \end{array}$$

When the log of a log is thus taken, a table giving at least one place more in the mantissa should be used than would be otherwise needed in order to protect the last place of figures in the result, as is shown in the above example, which is by no means an extreme case. If the quantity is a decimal fraction, the negative characteristic must be separately treated and the result subtracted from the result obtained with the mantissa.

Powers of Decimal Fractions.—Here it is necessary to notice that the logarithm of a decimal fraction is composed of two parts, one positive the other negative. Thus,

$$\log 0.06\,831 = \bar{2}.8345 \quad \text{or} \quad 8.8345 - 10.,$$

according to the notation chosen. In the first form (the one here recommended) the logarithm consists of a negative characteristic and a positive mantissa; in the second form it consists of a positive logarithm (both characteristic and mantissa) followed by a negative characteristic.

No new procedure is necessary, but it is essential to be consistent as to sign of each part, and to remember that both parts must be subjected to any operation performed upon either. Whenever the exponent has more than one figure, the negative characteristic and the positive mantissa should be separately multiplied by the index, and the former subtracted from the latter.

Example. — Desired $(0.00\,4716)^4$.

$$\begin{array}{lrcrcl}
\log 0.00\,4716 = & \bar{3}.6735 & \text{or} & 7.6735 - 10 & & \\
\text{Multiply by} & 4 & & 4 & & \\
\hline
 & \overline{10}.6940 & \text{or} & 30.6940 - 40, & \text{or} & 0.6940 - 10
\end{array}$$

$\therefore\ (0.00\,4716)^4$ = number corresponding, or antilog,
$= 4.943.10^{-10}$ or $0.00\,000\,000\,049\,43$.

Example. — Desired $(0.00\,4716)^{4.32}$.

$$\begin{array}{lrcrr}
\log 0.00\,4716 = & \bar{3}.6735 & \text{or} & 7.6735 - & 10 \\
 & 4.32 & & 4.32 & 4.32 \\
\hline
 & 2\,6940 & & 3\,06940 - & 43.2 \\
 & 20215 & & 230205 & \\
 & 1348 & & 15348 & \\
\hline
 & 2.9095 & & 33.14953 & \\
 & -\,12.96 & & -\,43.2 & \\
\hline
 & \overline{11}.9495 & & \overline{11}.9495 &
\end{array}$$

$\therefore\ (0.00\,4716)^{4.32} = 2.962 \cdot 10^{-11}$.

Roots of Decimal Fractions. — To extract the root of a decimal fraction, divide separately the negative characteristic and the positive mantissa by the index of the desired root. Subtract the former quotient from the latter. The difference will be the logarithm of the desired root. In other words, treat the characteristic separate from the logarithm. This procedure is always the safest to adopt when there is any doubt in the mind of the computer.

Example. — Desired $\sqrt[3]{0.06\,831}$.

$$\begin{array}{lr}
\log\ 0.06\,831 = & \bar{2}.8345 \\
\text{Dividing characteristic by } 3 = & -\,0.6667 \\
\text{Dividing mantissa by } 3 = & 0.2782 \\
\hline
\therefore\ \log \sqrt[3]{0.06831} = & \bar{1}.6115 \\
\sqrt[3]{0.06\,831} = \text{number corresponding} = & 4.088 \cdot 10^{-1}.
\end{array}$$

A method easily understood by inspection is this: Add to the negative characteristic of the logarithm a number equal to mr where r is the index of the desired root and m is a whole number large enough to make mr larger than or equal to the characteristic, in other words, large enough to extinguish the negative sign of the characteristic. Write this number with a negative sign after the logarithm. Then divide the whole by the index r. The quotient will be the logarithm of the desired root. Usually $m = 1$, *i.e.* $mr = r$.

Example. — Desired $\sqrt[3]{0.06\,831}$.

log 0.06 831 = $\bar{2}.8345$, adding and subtracting 3 gives $\underline{1.8345 - 3}$

dividing by 3 gives $0.6115 - 1 = \bar{1}.6115$

$\therefore\ \sqrt[3]{0.06\,831}$ = number corresponding = $4.088 \cdot 10^{-1}$ or 0.4088.

SQUARES AND SQUARE ROOTS.

It is common to give separate tables of squares and of square roots by which, respectively, the square or square root of any number may be taken out. Inspection, however, shows that the four-place table of square roots must occupy four pages, the first two containing the numbers from 1.0 to 10., the second two from 10. to 100.; the corresponding roots ranging from 1. to 10.; while a table of squares would occupy but two pages, containing numbers from 1.0 to 10., the squares ranging from 1. to 100. Hence the tabular differences in the table of square roots will be much smaller than in the table of squares. For this reason the table of squares may advantageously be dispensed with, and squares be taken out when desired from the table of square roots, as the numbers corresponding (antilogarithms) are taken out of a table of logarithms. The smallness of the differences makes interpolation easier and more rapid than in a table of squares, and this more than offsets the slight disadvantage of the reverse process of interpolation. The tables at page 30 give square roots to four places. As in the logarithm tables, the numbers in the body of the table are printed in stronger type wherever the second figure changes, in order to assist the eye in the reverse use of the table.

In the table of square roots the insertion of the extra section giving 1.00 to 1.10 to four places direct to avoid interpolation is not called for as in logarithms because the interpolation is very easy, and because the squares of terms of the form $(1 + a)$ are not frequently required.

To find the square of any number, factor the number as described in the "Notation by Powers of 10," page xv. Enter the body of the table with the first factor, and find the corresponding marginal reading and column heading which will be the first three figures of the square. Interpolate for the fourth figure. Square the factor $10^{\pm n}$, which makes it $10^{\pm 2n}$. Multiply together these two squares.

Example. — Desired the square of 34 850.

$$34\,850.^2 = (3.485 \cdot 10^4)^2 = 3.485^2 \cdot 10^{2\times 4}.$$

In the table 3.479 (the next smallest value to 3.485) stands in line 12, column 1. The tabular difference is 14, of which our difference (= 3.485 − 3.479 = 6) is $\frac{6}{14}$, or 0.4 (as may be seen at once from the interpolation table headed 14).

Hence $$34\,850.^2 = 12.14 \cdot 10^8 = 121\,400\,000.$$

Example. — Desired $(0.00\,049\,83)^2$.

$$(0.00\,049\,81)^2 = 4.981^2 \cdot 10^{-4\times 2} = 24.83 \cdot 10^{-8} = 0.00\,000\,024\,83.$$

To find the square root of any number, factor it as in the "Notation by Powers of 10," page xv, except that n **must be now an even number, while the first factor must have either one or two digits** preceding the decimal point, whichever may be necessary, in order to permit n to be made even.

Find, then, from the table, interpolating if necessary, the square root of the first factor, and multiply this by $10^{\pm n/2}$, the square root of the second factor; the product is the desired square root.

Example. — Desired the square root of 347.6.

$$(347.6)^{\frac{1}{2}} = (3.476 \cdot 10^2)^{\frac{1}{2}}.$$

In the table, line 3.4, column 7, gives $(3.47)^{\frac{1}{2}} = 1.863$
Interpolations by table gives 0.6 difference $= 1$

$$(3.476)^{\frac{1}{2}} = 1.864$$

$$\therefore\ (347.6)^{\frac{1}{2}} = 1.864 \cdot 10^1 = 18.64.$$

Example. — Desired $\sqrt{875\,200.}$

$$\sqrt{87.52 \cdot 10^4} = 9.355 \cdot 10^2 = 935.5.$$

Example. — Desired $(0.05643)^{\frac{1}{2}}$.

$$(5.643 \cdot 10^{-2})^{\frac{1}{2}} = 2.376 \cdot 10^{-1} = 0.2376.$$

Example. — Desired 0.00 006 784.

$$67.84 \cdot 10^{-6} = 8.236 \cdot 10^{-3} = 0.00\,823\,6.$$

RECIPROCALS.

This table, page 34, contains the reciprocals to four places of numbers from 1. to 9.99, and by interpolation from 1. to 9.999. The reciprocal to four places of any number from zero to infinity is, of course, one of these tabular values multiplied by a suitable power of 10. In this table, the reciprocals of numbers from 1. to 1.100 are given directly in the first ten lines to avoid interpolation in finding the reciprocals of terms of the form $(1 + a)$, where a is a small decimal fraction.

N.B. — The approximation $1/(1 + a) = (1 - a)$ approx., is frequently employed in computations to avoid dividing by $(1 + a)$. This useful approximation, however, must be used with caution. The error from its employment is a^2; that is, if $a = 0.1$, the error from multiplying by $(1 - a)$ instead of dividing by $(1 + a)$ is $0.1^2 = 0.01$, or 1 per cent; if $a = 0.03$, the error is $0.03^2 = 0.0009$, or nearly 0.1 per cent, and so on.

To find the reciprocal of a number, factor it by the "Notation by Powers of 10," page xv. Find the reciprocal of the first factor by the table and multiply this by the second factor $10^{\pm n}$ with the algebraic sign of its exponent reversed, *i.e.* by $10^{\mp n}$.

Example. — Desired 1/4486, *i.e.* $(4486)^{-1}$.

$$(4486.)^{-1} = (4.486 \cdot 10^3)^{-1} = (4.486.)^{-1} \cdot 10^{-3}.$$

In table, line 4.4, col. 8, $(4.48)^{-1} = 0.2232$

By difference table, 0.6 difference = -3

Reciprocal of $4.486 = 0.2229$

$$(4486.)^{-1} = (4.486)^{-1} \cdot 10^{-3} = 0.2229 \cdot 10^{-3} = 0.0002229.$$

Example. — Desired 1/0.00 327 5.

$$(0.00\ 327\ 5)^{-1} = (3.275)^{-1} \cdot (10^{-3})^{-1} = 0.3054 \cdot 10^3 = 305.4.$$

NATURAL SINES, COSINES, TANGENTS, AND COTANGENTS.

It is frequently convenient to have a table giving rough values of the natural trigonometric functions for use in preliminary, check, or approximate computations. The four-place tables of the above four functions cover all ordinary needs. Interpolation can be made in these tables to $0^\circ.01$ by the interpolation tables, or to $1'$ by mental computation, since the step between successive columns is $0^\circ.1$, or $6'$.

Example. — Desired the natural sine of 37°.75.

On line 37° under heading °.7 is 0.6115. To interpolate for the remaining figure we must add 0.5 of the tabular difference, which, by inspection, is 14. $0.5 \times 14 = 7$, as can be seen at once in the interpolation table 14 opposite the line 37°.

$$\therefore \sin 37^\circ.75 = 0.6115 + .0007 = 0.6122.$$

Example. — Desired the natural cotangent of 72°.28.

The cosines and cotangents read upward in the right-hand degree column.

Line 72°, column °.2, gives	0.3211
0.8 of difference (19), to be subtracted,	−15
∴ natural cotangent 72°.28	= 0.3196

All of the tables are used similarly.

LOGARITHMS OF SINES, COSINES, TANGENTS, AND COTANGENTS.

Two sets of tables are given, four-place and five-place, respectively. The four-place tables read to tenths and hundredths of degrees, and are convenient for general rough work and with instruments reading to tenths of a degree. The five-place tables give values to minutes direct. They are adapted to a very large part of the angular measurements ordinarily made in experimental work. Interpolation to half or quarter minutes is easily made in them mentally. A computer working closer than that would naturally employ the Vega, or other conveniently arranged six- or seven-place tables, dropping unnecessary places.

For the reasons stated under the discussion of logarithm tables, page xxviii, the negative characteristic is here retained instead of the more customary 9, with the appended − 10.

The use of the tables needs little explanation. The four-place tables are used precisely as the tables of the natural functions. In the five-place tables, for less than 45°, read the tables downwards, using the minute column at the left hand. For angles greater than 45°, read the tables upwards, using the minute column at the right hand of the table, and take care to employ the column-headings given at the foot of the table.

Examples. —

$$\text{Log sin } 31^\circ\ 22' = \bar{1}.71\ 643.$$

$$\text{Log tan } 54^\circ\ 46' = 0.15\ 101.$$

SLIDE WIRE RATIOS.

The increasing use of the slide wire in electrical measurements, notably in connection with physical chemistry, renders the table of values of

$$\frac{a}{1000 - a}$$

a decided convenience. The work to which the slide wire is thus for the most part employed demands but four-place tables.

To use the table, let a be the reading in millimetres on the slide wire, which is supposed to be one metre long with millimetre subdivision, or of any other length with division into thousandths. Find the first two figures of a in the first column, and run out on this line to the column headed with the third figure. The number there found will give the value of the above fraction, which is the ratio of the length of one section of the wire to the other. If a contains a fourth figure, that is, if read to tenths of a millimetre, interpolation must be made in the usual manner.

Example. — The slider reads 415.6 millimetres.

In line 41 under 5 is found 0.7094. Interpolating by adding 0.6 of the tabular difference 29, gives

$$x = 0.7094 + 0.6 \times 29$$
$$= 0.7094 + .0017 = 0.7111.$$

DEFINITIONS AND EXPLANATIONS

UNDERLYING THE COMPUTATION RULES.

The following statements call for the attention of those only who find unfamilar terms in the foregoing rules.

A Digit is any one of the ten characters 1, 2, 3, 4, 5, 6, 7, 8, 9, 0.

A Significant Figure is any digit used to denote or signify the amount of the quantity in the place in which it stands. Thus zero may be a significant figure when it is written not merely to locate the decimal point, but to indicate that the quantity in the place in which it stands is known to be nearer to zero than to any other digit.

For example, if a distance has been measured to the nearest fiftieth of an inch and found to be 205.46 inches, all five of the figures, including the zero, are significant. And similarly if the measurement had shown the distance to be nearer to 205.40 than to 205.41 or to 205.39 the *final zero* would be also significant, and *should invariably be retained*, since its presence serves the most useful purpose of showing that this place of figures had been measured as well as the rest. If in such a case the quantity had been written 205.4 instead of 205.40, the inference would be drawn either that the hundredths of an inch had not been measured, or that the person who wrote the number was ignorant or careless of the proper numerical usage. Failure to follow this simple rule is a common source of annoyance and uncertainty.

A zero when used merely to enable the decimal point to be retained is of course not a significant figure in the above sense. *E.g.* If the distance were measured as 286. centimetres within ± 1 centimetre, it might be retained as 2.86 metres, or 0.00 286 kilometres, or 2860. millimetres. In this case neither of the zeros would be significant.

From this last example it is obvious that when the first zero, or zeros, precede the decimal point, they fail to indicate intrinsically whether or not they are significant. *E.g.* The number 2860. millimetres or 28 600. millimetres, standing apart from explanatory context, would afford no clew to whether the last place, or the last two places, had been measured. In writing such a number, therefore, whenever it is desirable to convey this information some statement of the reliability of the quantity must be appended. This is usually done by writing after the number $\pm a$ where a is a number, of two significant figures only, representing the estimated measure of the accuracy or unreliability of the result. (See later.)

Places of Figures are the places in which figures stand in the number as actually written. Places of significant figures are those in which significant figures stand. These two terms being merely, although not actually, identical, are often used interchangeably.

Example.—426 018. has six places of figures and of significant figures 3 479 100. has seven places of figures, but whether its number of places of significant figures is more than five is indeterminate. 0.02 7680 has seven places of figures with five places of significant figures. $2.76\,80 \cdot 10^{-2}$ has five places of figures and five significant figures.

Places of Decimals.—These, of course, are the places following the decimal point as the number happens to be written. *E.g.* 0.02 768 0 has six decimal places. $2.76\,80 \cdot 10^{-2}$ has four decimal places, although its magnitude is the same as that of the preceding quantity.

From the foregoing six examples it will be seen that the number of decimal places and the number of places of significant figures have no necessary mutual relation.

Accuracy; Reliability.—To know the exact accuracy of a given quantity it would be necessary, of course, to know the true value of that quantity. In the case of a few mathematical constants (such as $\pi = 3.14\,159\,265 \cdots$) the true value *is* known, at least far beyond ordinary requirements. But in case of all measurements the same is obviously not true, for if the true value were known, measurements would be unnecessary. *Some approximate expression* of the accuracy of a measured result can, however, usually be obtained, and is necessary. Sometimes this is afforded by a knowledge of the instrument used and the degree of care employed. Thus, suppose the distance of about 3 feet $6\frac{3}{64}$ inches = 3.5039 feet between two marks to have been carefully once measured with a good foot-rule, it could safely be assumed from our previous knowledge of such work that if a series of these measurements were undertaken, the results would vary from their mean by less than $\pm\frac{1}{32}$ of an inch; also, that the error in the rule itself, and from other unavoidable or unavoided sources, would not on the average materially increase the error of measurement. Then $\pm\frac{1}{32}$ inch $= \pm 0.0026$ ft. would be taken as the *estimated measure of accuracy of the result.* Instead of expressing the accuracy in units, *e.g.* inches or feet, it is usually more convenient or intelligible to express it as a fraction; or, better still, in percentage. Thus, the foregoing will be

$$\pm\frac{0.0026}{3.5} = \pm 0.00\,074, \textit{ i.e. } \frac{74}{100\,000}, \text{ or } 100\left(\pm\frac{0.0026}{3.5}\right) = \pm 0.074 \text{ per cent.}$$

If in this example the reliability had not been estimated at $\frac{1}{32}$ inch, but if several measurements had been made, and these had been found upon inspection to deviate from their mean by about $\frac{1}{32}$ of an inch, then, other things being the same, the measure of accuracy of any single measurement taken without knowledge of the others would be regarded as $\pm\frac{1}{32}$ inch, or ± 0.074 per cent.

Mean; Average; Deviation Measure.—When the result is the arithmetical mean or average of several separate measurements of the same quantity, its reliability or accuracy is taken to be in proportion to the square root of the number of such observations. Thus, in the last preceding example, if there had been $n = 9$ single observations made, the measure of accuracy of the mean of these would be

$$\pm\frac{1}{32} \div \sqrt{n} = \pm\frac{1}{32}\,\frac{1}{\sqrt{9}} = \pm\frac{1}{96} = \pm 0.010 \text{ inch, or } \pm 0.025 \text{ per cent.}$$

The differences of the single observations taken under like conditions from their mean will be called deviations, and their numerical average (*i.e.* their sum omitting their algebraic sign, divided by their number) will be called the average deviation of the single observation from the mean, and will be denoted by *ad*. This quantity divided by the square root of *n* would be called the average deviation of the mean, and will be denoted by *AD*. It is the average amount by which any one such mean would be found to deviate from the average of a number of such means all taken under like conditions. The term *deviation measure* will be used in referring to either of these quantities.

Briefly, then, the meaning of the terms may be summarized as follows: By the statement that the *accuracy* or *reliability* of a result is of a quoted amount is meant that the deviation measure of this result is estimated or found to be of this stated amount, and that so far as this can be discovered by inspection, no other sources of error exist which affect the result by an amount sensible as compared with this.*

It is essential to bear in mind in connection with all quantities that are the result of measurement that no *absolute* numerical expression of the accuracy or error is possible; that any expression which is given is usually merely a deviation measure; that is, an approximate average value of the effect of the variable parts of the errors, accompanied by an assurance, expressed or implied, that a study of the discoverable sources of error of the process has been made, and that these have been corrected for or found to be negligible compared to the deviation measure.

In specifying the accuracy desired in the result, it must be understood that merely the converse of this is meant; or, at least, that only the converse is possible of attainment.

Rules for Significant Figures. — The rules are so framed that, barring mistakes, the greatest possible computation error entering into the result of any ordinary computation (*e.g.* one involving a total of not much exceeding 20 component numbers, steps, or operations, where a rejection error may occur) shall not be sensible compared with the errors of the measurements or data, or shall not sensibly affect the accuracy of the result. They are, therefore, *safe* rules in the worst possible cases. But in order to be so they are necessarily more than sufficiently stringent for some classes of comparatively rough work, where the *infrequent* undetected entrance of a computation error two to four times as large as the experimental error would be permissible. For such work one less place of figures may be used, but when the rules are thus relaxed the possible consequence should be borne in mind, and special scrutiny applied to the various stages of the computation, special attention being directed to quantities beginning with 1 or 2.

Rejection Error. — Whenever it becomes necessary to throw off places of figures, a "rejection error" may enter into the result; *e.g.* suppose that for any reason the last two figures are to be rejected in

24 375 291,

making it 24 375 300.

The rejection error in the new form is evidently + 9.

In rejection, the last figure retained is always to be increased by 1 when the rejected figure next it is 5 or over, but remains unchanged if that figure is less than 5.

* For a more extended discussion of this and allied subjects see the author's "Precision of Measurements."

Thus, calling the last place retained the rth place, the limits of the rejection error entering into that place are $+0.5$ and -0.5, and as all amounts between these limits are equally likely to occur, the average rejection error in the long run will be 0.25 in the rth place.

Law and Amount of Accumulated Rejection Error.—Let the last place of significant figures retained in a number, *e.g.* the fourth, fifth, etc., be called the rth place. Then the error entering from rejection of figures beyond the rth will be at most ± 5 in the $(r+1)$ place, and any error between these limits, $+5$ and -5 in the $(r+1)$ place, will be equally likely to occur in any given case, and therefore will be of equal frequency in the long run. The average rejection error in any considerable number of rejections will therefore be ± 2.5 in the $(r+1)$ place. If then in direct processes of multiplication, division, evolution, or involution (separate or combined) each factor, product, and quotient in the operation be carried out to the same number r of places, what will be the accumulated fractional rejection error if n such rejections are made during the entire operation? This error we shall call for brevity the computation error, or simply the rejection error. This question might easily be answered in a form giving the *average* accumulated error, but we are at present concerned chiefly with the *maximum* possible error, since our object is to frame rules which will reduce the worst possible computation error to negligible dimensions. The maximum computation error would arise when every single rejection error was 5 in the $(r+1)$ place, and all had the same sign; and this would be the greatest fraction of the final result when that result and all the factors (and therefore all the intermediate products, quotients, etc.) began with 1 and had 0 in the other places, *i.e.* were each 10^{r-1} with the decimal point wherever it might happen to stand. If there were n of these factors, products, quotients, etc., at which rejections were made, the maximum possible computation error in the result would, therefore, be $n \times 5$ in the $(r+1)$ place, and the fractional error would be $5n/10^r$. Since this maximum error would be exceedingly rare unless n were very small, and any approaching one-half of it would be very rare, we may properly assume that our rules will be sufficiently stringent if we allow this maximum error to have the same magnitude as the desired accuracy in the result as expressed on the basis of the precision measure, or deviation measure, explained at page xlii. To determine most simply what number r of places this limitation would call for in the processes under consideration, let us take specific cases. Suppose the work is desired to possess an accuracy of 1 per cent. Then $5n/10^r$ must be equal to or less than 1 per cent; *i.e.* $5n \leqq 1/100$. Hence, we have to solve

$$\frac{5n}{10^r} = \frac{1}{100}.$$

By inspection, if $r = 3,\ 5n = 10,\ \therefore n = 2,$
if $r = 4,\ 5n = 100,\ \therefore n = 20.$

But obviously n will almost never be as small as 2, and rarely as large as 20, lying with greatest frequency between 5 and 10 and averaging below 7. This will give a maximum error of 0.0035 or $\frac{1}{3}$ per cent with $n = 7$, $r = 4$, which would be insignificant, and rising to 1 per cent only when $n = 20$. Hence, in work of multiplication, division, etc., where 1 per cent accuracy, or a little better, is desired (remembering that by this we mean only a *deviation measure* of 1 per cent) $r = 4$ will be an entirely safe but not an excessive value; that is, the retention of four significant figures throughout will insure entirely sufficient freedom from computation error in every case when the number of rejections is less than the very unusual total of about 20, and fewer places would not be warranted. Similarly five places will suffice for work to 0.1 per cent, *i.e.* better than 1 per cent, but not much better than 0.1 per cent; six places for work of 0.01 per cent, and so on.

As to relaxation of the rules in special cases, it is evident that but little can be safely done. The maximum error with $n = 7$, $r = 4$, will be 0.35 per cent, and an error of $\frac{1}{4}$ of this, say, of 0.1 per cent, will not be of very uncommon occurrence. In fact, with only two rejections the error might be 0.1 per cent. Hence, four places cannot be considered as safe to 0.1 per cent, even for short computations. Four places might properly enough be used in short computations up to $\frac{1}{2}$ per cent. Similar statements of course hold for the other rules.

When **logarithms** are used for multiplication, division, etc., tables should be used giving the mantissa to the same number of places of figures as required by the foregoing rules for direct multiplication and division. Hence, one would employ four-place tables for 1 per cent work, five-place tables for 0.1 per cent work, and so on. The numbers, antilogs, and mantissæ should all be carried out to the same number of places. This conforms to the customary and only convenient practice.

This rule is based primarily on the fact, next to be shown, that under them the maximum computation error in the use of logarithms arises chiefly from the rejection in the numbers and antilogs themselves and not sensibly from the rejected places in the logarithms. In logarithms, a change of the rth figure by 1 produces the same fractional error in the antilog whatever its value, viz. $2.4/10^r$, as may be seen easily by inspection of tables. Hence, as the maximum rejection error in the tabular value of a logarithm is 5 in the $(r + 1)$ place of the mantissæ, which may be doubled by the process of interpolating, the maximum fractional error in a result due to the maximum rejection error in any mantissa is $2.4/10^r$. But if only r places are retained in the number or antilog, the maximum error in it due to rejection of its further places is $5/10^r$, compared to which $2.4/10^r$ is negligible. Hence, as the number of rejections from number and antilog together is usually about the same as from mantissæ, the accumulated error will be due almost wholly to the rejections from the numbers and antilogs. And of these rejections there will ordinarily be about as many if the computation is carried out by logarithms as if by direct multiplication or division. The above rule is thus justified.

In **addition or subtraction** the maximum rejection error will be obviously $5n$ in the $(r + 1)$ place. Under the above stated rule that the weakest quantity shall be carried to four significant figures (or two uncertain places) for 1 per cent work, the smallest value of the deviation measure or uncertainty of the result will be 10 in the rth place, which is 100 in the $(r + 1)$ place. Hence,

$$5n \gtrless 100. \quad \therefore \; n = 20.$$

The maximum accumulated error would then attain the size of the smallest deviation measure, *i.e.* the worst possible case would occur, only when the number of rejections was as great as twenty. Hence, the rule of four places for 1 per cent work, five for 0.1 per cent work, and so on, as before given, is sufficient.

TABLES

FOUR PLACE LOGARITHMS.

No.	0	1	2	3	4	5	6	7	8	9	INTERPOLATION TABLES			
1.00	.0000	.0004	.0009	.0013	.0017	.0022	.0026	.0030	.0035	.0039	**38**	**36**	**34**	**32**
.01	0043	0048	0052	0056	0060	0065	0069	0073	0077	0082	4	4	3	3
.02	0086	0090	0095	0099	0103	0107	0111	0116	0120	0124	8	7	7	6
.03	0128	0133	0137	0141	0145	0149	0154	0158	0162	0166	11	11	10	10
.04	0170	0175	0179	0183	0187	0191	0195	0199	0204	0208	15	14	14	13
1.05	.0212	.0216	.0220	.0224	.0228	.0233	.0237	.0241	.0245	.0249	19	18	17	16
.06	0253	0257	0261	0265	0269	0273	0278	0282	0286	0290	23	22	20	19
.07	0294	0298	0302	0306	0310	0314	0318	0322	0326	0330	27	25	24	22
.08	0334	0338	0342	0346	0350	0354	0358	0362	0366	0370	30	29	27	26
.09	0374	0378	0382	0386	0390	0394	0398	0402	0406	0410	34	32	31	29
1.0	.0000	.0043	.0086	.0128	.0170	.0212	.0253	.0294	.0334	.0374	**30**	**28**	**26**	**24**
.1	0414	0453	0492	0531	0569	0607	0645	0682	0719	0755	3	3	3	2
.2	0792	0828	0864	0899	0934	0969	**1004**	1038	1072	1106	6	6	5	5
.3	1139	1173	1206	1239	1271	1303	1335	1367	1399	1430	9	8	8	7
.4	1461	1492	1523	1553	1584	1614	1644	1673	1703	1732	12	11	10	10
1.5	.1761	.1790	.1818	.1847	.1875	.1903	.1931	.1959	.1987	**.2014**	15	14	13	12
.6	2041	2068	2095	2122	2148	2175	2201	2227	2253	2279	18	17	16	14
.7	2304	2330	2355	2380	2405	2430	2455	2480	2504	2529	21	20	18	17
.8	2553	2577	2601	2625	2648	2672	2695	2718	2742	2765	24	22	21	19
.9	2788	2810	2833	2856	2878	2900	2923	2945	2967	2989	27	25	23	22
2.0	**.3010**	.3032	.3054	.3075	.3096	.3118	.3139	.3160	.3181	.3201	**22**	**20**	**18**	**16**
.1	3222	3243	3263	3284	3304	3324	3345	3365	3385	3404	2	2	2	2
.2	3424	3444	3464	3483	3502	3522	3541	3560	3579	3598	4	4	4	3
.3	3617	3636	3655	3674	3692	3711	3729	3747	3766	3784	7	6	5	5
.4	3802	3820	3838	3856	3874	3892	3909	3927	3945	3962	9	8	7	6
2.5	.3979	.3997	**.4014**	.4031	.4048	.4065	.4082	.4099	.4116	.4133	11	10	9	8
.6	4150	4166	4183	4200	4216	4232	4249	4265	4281	4298	13	12	11	10
.7	4314	4330	4346	4362	4378	4393	4409	4425	4440	4456	15	14	13	11
.8	4472	4487	4502	4518	4533	4548	4564	4579	4594	4609	18	16	14	13
.9	4624	4639	4654	4669	4683	4698	4713	4728	4742	4757	20	18	16	14
3.0	.4771	.4786	.4800	.4814	.4829	.4843	.4857	.4871	.4886	.4900	**15**	**14**	**13**	**12**
.1	4914	4928	4942	4955	4969	4983	4997	**5011**	5024	5038	2	1	1	1
.2	5051	5065	5079	5092	5105	5119	5132	5145	5159	5172	3	3	3	2
.3	5185	5198	5211	5224	5237	5250	5263	5276	5289	5302	5	4	4	4
.4	5315	5328	5340	5353	5366	5378	5391	5403	5416	5428	6	6	5	5
3.5	.5441	.5453	.5465	.5478	.5490	.5502	.5515	.5527	.5539	.5551	8	7	7	6
.6	5563	5575	5587	5599	5611	5623	5635	5647	5658	5670	9	8	8	7
.7	5682	5694	5705	5717	5729	5740	5752	5763	5775	5786	11	10	9	8
.8	5798	5809	5821	5832	5843	5855	5866	5877	5888	5899	12	11	10	10
.9	5911	5922	5933	5944	5955	5966	5977	5988	5999	**6010**	14	13	12	11
4.0	.6021	.6031	.6042	.6053	.6064	.6075	.6085	.6096	.6107	.6117	**11**	**10**	**9**	**8**
.1	6128	6138	6149	6160	6170	6180	6191	6201	6212	6222	1	1	1	1
.2	6232	6243	6253	6263	6274	6284	6294	6304	6314	6325	2	2	2	2
.3	6335	6345	6355	6365	6375	6385	6395	6405	6415	6425	3	3	3	2
.4	6435	6444	6454	6464	6474	6484	6493	6503	6513	6522	4	4	4	3
4.5	.6532	.6542	.6551	.6561	.6571	.6580	.6590	.6599	.6609	.6618	6	5	5	4
.6	6628	6637	6646	6656	6665	6675	6684	6693	6702	6712	7	6	5	5
.7	6721	6730	6739	6749	6758	6767	6776	6785	6794	6803	8	7	6	6
.8	6812	6821	6830	6839	6848	6857	6866	6875	6884	6893	9	8	7	6
.9	6902	6911	6920	6928	6937	6946	6955	6964	6972	6981	10	9	8	7

FOUR PLACE LOGARITHMS.

No.	0	1	2	3	4	5	6	7	8	9	Interpolation Tables.		
5.0	.6990	.6998	.7007	.7016	.7024	.7033	.7042	.7050	.7059	.7067	**9**	**8**	**7**
.1	7076	7084	7093	7101	7110	7118	7126	7135	7143	7152	1	1	1
.2	7160	7168	7177	7185	7193	7202	7210	7218	7226	7235	2	2	1
.3	7243	7251	7259	7267	7275	7284	7292	7300	7308	7316	3	2	2
.4	7324	7332	7340	7348	7356	7364	7372	7380	7388	7396	4	3	3
5.5	.7404	.7412	.7419	.7427	.7435	.7443	.7451	.7459	.7466	.7474	5	4	4
.6	7482	7490	7497	7505	7513	7520	7528	7536	7543	7551	5	5	4
.7	7559	7566	7574	7582	7589	7597	7604	7612	7619	7627	6	6	5
.8	7634	7642	7649	7657	7664	7672	7679	7686	7694	7701	7	6	6
.9	7709	7716	7723	7731	7738	7745	7752	7760	7767	7774	8	7	6
6.0	.7782	.7789	.7796	.7803	.7810	.7818	.7825	.7832	.7839	.7846	**7**		**6**
.1	7853	7860	7868	7875	7882	7889	7896	7903	7910	7917	1		1
.2	7924	7931	7938	7945	7952	7959	7966	7973	7980	7987	1		1
.3	7993	**8000**	8007	8014	8021	8028	8035	8041	8048	8055	2		2
.4	8062	8069	8075	8082	8089	8096	8102	8109	8116	8122	3		2
6.5	.8129	.8136	.8142	.8149	.8156	.8162	.8169	.8176	.8182	.8189	4		3
.6	8195	8202	8209	8215	8222	8228	8235	8241	8248	8254	4		4
.7	8261	8267	8274	8280	8287	8293	8299	8306	8312	8319	5		4
.8	8325	8331	8338	8344	8351	8357	8363	8370	8376	8382	6		5
.9	8388	8395	8401	8407	8414	8420	8426	8432	8439	8445	6		5
7.0	.8451	.8457	.8463	.8470	.8476	.8482	.8488	.8494	.8500	.8506	**6**		**5**
.1	8513	8519	8525	8531	8537	8543	8549	8555	8561	8567	1		1
.2	8573	8579	8585	8591	8597	8603	8609	8615	8621	8627	1		1
.3	8633	8639	8645	8651	8657	8663	8669	8675	8681	8686	2		2
.4	8692	8698	8704	8710	8716	8722	8727	8733	8739	8745	2		2
7.5	.8751	.8756	.8762	.8768	.8774	.8779	.8785	.8791	.8797	.8802	3		3
.6	8808	8814	8820	8825	8831	8837	8842	8848	8854	8859	4		3
.7	8865	8871	8876	8882	8887	8893	8899	8904	8910	8915	4		4
.8	8921	8927	8932	8938	8943	8949	8954	8960	8965	8971	5		4
.9	8976	8982	8987	8993	8998	**9004**	9009	9015	9020	9025	5		5
8.0	.9031	.9036	.9042	.9047	.9053	.9058	.9063	.9069	.9074	.9079	**6**		**5**
.1	9085	9090	9096	9101	9106	9112	9117	9122	9128	9133	1		1
.2	9138	9143	9149	9154	9159	9165	9170	9175	9180	9186	1		1
.3	9191	9196	9201	9206	9212	9217	9222	9227	9232	9238	2		2
.4	9243	9248	9253	9258	9263	9269	9274	9279	9284	9289	2		2
8.5	.9294	.9299	.9304	.9309	.9315	.9320	.9325	.9330	.9335	.9340	3		3
.6	9345	9350	9355	9360	9365	9370	9375	9380	9385	9390	4		3
.7	9395	9400	9405	9410	9415	9420	9425	9430	9435	9440	4		4
.8	9445	9450	9455	9460	9465	9469	9474	9479	9484	9489	5		4
.9	9494	9499	9504	9509	9513	9518	9523	9528	9533	9538	5		5
9.0	.9542	.9547	.9552	.9557	.9562	.9566	.9571	.9576	.9581	.9586	**5**		**4**
.1	9590	9595	9600	9605	9609	9614	9619	9624	9628	9633	1		0
.2	9638	9643	9647	9652	9657	9661	9666	9671	9675	9680	1		1
.3	9685	9689	9694	9699	9703	9708	9713	9717	9722	9727	2		1
.4	9731	9736	9741	9745	9750	9754	9759	9763	9768	9773	2		2
9.5	.9777	.9782	.9786	.9791	.9795	.9800	.9805	.9809	.9814	.9818	3		2
.6	9823	9827	9832	9836	9841	9845	9850	9854	9859	9863	3		2
.7	9868	9872	9877	9881	9886	9890	9894	9899	9903	9908	4		3
.8	9912	9917	9921	9926	9930	9934	9939	9943	9948	9952	4		3
.9	9956	9961	9965	9969	9974	9978	9983	9987	9991	9996	5		4

FOUR PLACE ANTILOGARITHMS.

Mant.	0	1	2	3	4	5	6	7	8	9	Interpola. Tables.			
	THOUSANDTHS.													
.00	1.000	1.002	1.005	1.007	1.009	1.012	1.014	1.016	1.019	1.021	**2**	**3**		
.01	1.023	1.026	1.028	1.030	1.033	1.035	1.038	1.040	1.042	1.045	0	0		
.02	1.047	1.050	1.052	1.054	1.057	1.059	1.062	1.064	1.067	1.069	0	1		
.03	1.072	1.074	1.076	1.079	1.081	1.084	1.086	1.089	1.091	1.094	1	1		
.04	1.096	1.099	1.102	1.104	1.107	1.109	1.112	1.114	1.117	1.119	1	1		
.05	1.122	1.125	1.127	1.130	1.132	1.135	1.138	1.140	1.143	1.146	1	2		
.06	1.148	1.151	1.153	1.156	1.159	1.161	1.164	1.167	1.169	1.172	1	2		
.07	1.175	1.178	1.180	1.183	1.186	1.189	1.191	1.194	1.197	1.199	1	2		
.08	1.202	1.205	1.208	1.211	1.213	1.216	1.219	1.222	1.225	1.227	2	2		
.09	1.230	1.233	1.236	1.239	1.242	1.245	1.247	1.250	1.253	1.256	2	3		
.10	1.259	1.262	1.265	1.268	1.271	1.274	1.276	1.279	1.282	1.285	**3**	**4**		
.11	1.288	1.291	1.294	1.297	1.300	1.303	1.306	1.309	1.312	1.315	0	0		
.12	1.318	1.321	1.324	1.327	1.330	1.334	1.337	1.340	1.343	1.346	1	1		
.13	1.349	1.352	1.355	1.358	1.361	1.365	1.368	1.371	1.374	1.377	1	1		
.14	1.380	1.384	1.387	1.390	1.393	1.396	1.400	1.403	1.406	1.409	1	2		
.15	1.413	1.416	1.419	1.422	1.426	1.429	1.432	1.435	1.439	1.442	2	2		
.16	1.445	1.449	1.452	1.455	1.459	1.462	1.466	1.469	1.472	1.476	2	2		
.17	1.479	1.483	1.486	1.489	1.493	1.496	1.500	1.503	1.507	1.510	2	3		
.18	1.514	1.517	1.521	1.524	1.528	1.531	1.535	1.538	1.542	1.545	2	3		
.19	1.549	1.552	1.556	1.560	1.563	1.567	1.570	1.574	1.578	1.581	3	4		
.20	1.585	1.589	1.592	1.596	1.600	1.603	1.607	1.611	1.614	1.618	**3**	**4**	**5**	
.21	1.622	1.626	1.629	1.633	1.637	1.641	1.644	1.648	1.652	1.656	0	0	1	
.22	1.660	1.663	1.667	1.671	1.675	1.679	1.683	1.687	1.690	1.694	1	1	1	
.23	1.698	1.702	1.706	1.710	1.714	1.718	1.722	1.726	1.730	1.734	1	1	2	
.24	1.738	1.742	1.746	1.750	1.754	1.758	1.762	1.766	1.770	1.774	1	2	2	
.25	1.778	1.782	1.786	1.791	1.795	1.799	1.803	1.807	1.811	1.816	2	2	3	
.26	1.820	1.824	1.828	1.832	1.837	1.841	1.845	1.849	1.854	1.858	2	2	3	
.27	1.862	1.866	1.871	1.875	1.879	1.884	1.888	1.892	1.897	1.901	2	3	4	
.28	1.905	1.910	1.914	1.919	1.923	1.928	1.932	1.936	1.941	1.945	2	3	4	
.29	1.950	1.954	1.959	1.963	1.968	1.972	1.977	1.982	1.986	1.991	3	4	5	
.30	1.995	2.000	2.004	2.009	2.014	2.018	2.023	2.028	2.032	2.037	**4**	**5**	**6**	
.31	2.042	2.046	2.051	2.056	2.061	2.065	2.070	2.075	2.080	2.084	0	1	1	
.32	2.089	2.094	2.099	2.104	2.109	2.113	2.118	2.123	2.128	2.133	1	1	1	
.33	2.138	2.143	2.148	2.153	2.158	2.163	2.168	2.173	2.178	2.183	1	2	2	
.34	2.188	2.193	2.198	2.203	2.208	2.213	2.218	2.223	2.228	2.234	2	2	2	
.35	2.239	2.244	2.249	2.254	2.259	2.265	2.270	2.275	2.280	2.286	2	3	3	
.36	2.291	2.296	2.301	2.307	2.312	2.317	2.323	2.328	2.333	2.339	2	3	4	
.37	2.344	2.350	2.355	2.360	2.366	2.371	2.377	2.382	2.388	2.393	3	4	4	
.38	2.399	2.404	2.410	2.415	2.421	2.427	2.432	2.438	2.443	2.449	3	4	5	
.39	2.455	2.460	2.466	2.472	2.477	2.483	2.489	2.495	2.500	2.506	4	5	5	
.40	2.512	2.518	2.523	2.529	2.535	2.541	2.547	2.553	2.559	2.564	**5**	**6**	**7**	**8**
.41	2.570	2.576	2.582	2.588	2.594	2.600	2.606	2.612	2.618	2.624	1	1	1	1
.42	2.630	2.636	2.642	2.649	2.655	2.661	2.667	2.673	2.679	2.685	1	1	1	2
.43	2.692	2.698	2.704	2.710	2.716	2.723	2.729	2.735	2.742	2.748	2	2	2	2
.44	2.754	2.761	2.767	2.773	2.780	2.786	2.793	2.799	2.805	2.812	2	2	3	3
.45	2.818	2.825	2.831	2.838	2.844	2.851	2.858	2.864	2.871	2.877	3	3	4	4
.46	2.884	2.891	2.897	2.904	2.911	2.917	2.924	2.931	2.938	2.944	3	4	4	5
.47	2.951	2.958	2.965	2.972	2.979	2.985	2.992	2.999	3.006	3.013	4	4	5	6
.48	3.020	3.027	3.034	3.041	3.048	3.055	3.062	3.069	3.076	3.083	4	5	6	6
.49	3.090	3.097	3.105	3.112	3.119	3.126	3.133	3.141	3.148	3.155	5	5	6	7

Mant.	0	1	2	3	4	5	6	7	8	9	Interpola. Tables.			
					THOUSANDTHS.									
.50	3.162	3.170	3.177	3.184	3.192	3.199	3.206	3.214	3.221	3.228	**7**	**8**	**9**	
.51	3.236	3.243	3.251	3.258	3.266	3.273	3.281	3.289	3.296	3.304	1	1	1	
.52	3.311	3.319	3.327	3.334	3.342	3.350	3.357	3.365	3.373	3.381	1	2	2	
.53	3.388	3.396	3.404	3.412	3.420	3.428	3.436	3.443	3.451	3.459	2	2	3	
.54	3.467	3.475	3.483	3.491	3.499	3.508	3.516	3.524	3.532	3.540	3	3	4	
.55	3.548	3.556	3.565	3.573	3.581	3.589	3.597	3.606	3.614	3.622	4	4	5	
.56	3.631	3.639	3.648	3.656	3.664	3.673	3.681	3.690	3.698	3.707	4	5	5	
.57	3.715	3.724	3.733	3.741	3.750	3.758	3.767	3.776	3.784	3.793	5	6	6	
.58	3.802	3.811	3.819	3.828	3.837	3.846	3.855	3.864	3.873	3.882	6	6	7	
.59	3.890	3.899	3.908	3.917	3.926	3.936	3.945	3.954	3.963	3.972	6	7	8	
.60	3.981	3.990	3.999	4.009	4.018	4.027	4.036	4.046	4.055	4.064	**9**	**10**	**11**	**12**
.61	4.074	4.083	4.093	4.102	4.111	4.121	4.130	4.140	4.150	4.159	1	1	1	1
.62	4.169	4.178	4.188	4.198	4.207	4.217	4.227	4.236	4.246	4.256	2	2	2	2
.63	4.266	4.276	4.285	4.295	4.305	4.315	4.325	4.335	4.345	4.355	3	3	3	4
.64	4.365	4.375	4.385	4.395	4.406	4.416	4.426	4.436	4.446	4.457	4	4	4	5
.65	4.467	4.477	4.487	4.498	4.508	4.519	4.529	4.539	4.550	4.560	5	5	6	6
.66	4.571	4.581	4.592	4.603	4.613	4.624	4.634	4.645	4.656	4.667	5	6	7	7
.67	4.677	4.688	4.699	4.710	4.721	4.732	4.742	4.753	4.764	4.775	6	7	8	8
.68	4.786	4.797	4.808	4.819	4.831	4.842	4.853	4.864	4.875	4.887	7	8	9	10
.69	4.898	4.909	4.920	4.932	4.943	4.955	4.966	4.977	4.989	5.000	8	9	10	11
.70	5.012	5.023	5.035	5.047	5.058	5.070	5.082	5.093	5.105	5.117	**12**	**13**	**14**	**15**
.71	5.129	5.140	5.152	5.164	5.176	5.188	5.200	5.212	5.224	5.236	1	1	1	2
.72	5.248	5.260	5.272	5.284	5.297	5.309	5.321	5.333	5.346	5.358	2	3	3	3
.73	5.370	5.383	5.395	5.408	5.420	5.433	5.445	5.458	5.470	5.483	4	4	4	5
.74	5.495	5.508	5.521	5.534	5.546	5.559	5.572	5.585	5.598	5.610	5	5	6	6
.75	5.623	5.636	5.649	5.662	5.675	5.689	5.702	5.715	5.728	5.741	6	7	7	8
.76	5.754	5.768	5.781	5.794	5.808	5.821	5.834	5.848	5.861	5.875	7	8	8	9
.77	5.888	5.902	5.916	5.929	5.943	5.957	5.970	5.984	5.998	6.012	8	9	10	11
.78	6.026	6.039	6.053	6.067	6.081	6.095	6.109	6.124	6.138	6.152	10	10	11	12
.79	6.166	6.180	6.194	6.209	6.223	6.237	6.252	6.266	6.281	6.295	11	12	13	14
.80	6.310	6.324	6.339	6.353	6.368	6.383	6.397	6.412	6.427	6.442	**16**	**17**	**18**	**19**
.81	6.457	6.471	6.486	6.501	6.516	6.531	6.546	6.561	6.577	6.592	2	2	2	2
.82	6.607	6.622	6.637	6.653	6.668	6.683	6.699	6.714	6.730	6.745	3	3	4	4
.83	6.761	6.776	6.792	6.808	6.823	6.839	6.855	6.871	6.887	6.902	5	5	5	6
.84	6.918	6.934	6.950	6.966	6.982	6.998	7.015	7.031	7.047	7.063	6	7	7	8
.85	7.079	7.096	7.112	7.129	7.145	7.161	7.178	7.194	7.211	7.228	8	9	9	10
.86	7.244	7.261	7.278	7.295	7.311	7.328	7.345	7.362	7.379	7.396	10	10	11	11
.87	7.413	7.430	7.447	7.464	7.482	7.499	7.516	7.534	7.551	7.568	11	12	13	13
.88	7.586	7.603	7.621	7.638	7.656	7.674	7.691	7.709	7.727	7.745	13	14	14	15
.89	7.762	7.780	7.798	7.816	7.834	7.852	7.870	7.889	7.907	7.925	14	15	16	17
.90	7.943	7.962	7.980	7.998	8.017	8.035	8.054	8.072	8.091	8.110	**20**	**21**	**22**	**23**
.91	8.128	8.147	8.166	8.185	8.204	8.222	8.241	8.260	8.279	8.299	2	2	2	2
.92	8.318	8.337	8.356	8.375	8.395	8.414	8.433	8.453	8.472	8.492	4	4	4	5
.93	8.511	8.531	8.551	8.570	8.590	8.610	8.630	8.650	8.670	8.690	6	6	7	7
.94	8.710	8.730	8.750	8.770	8.790	8.810	8.831	8.851	8.872	8.892	8	8	9	9
.95	8.913	8.933	8.954	8.974	8.995	9.016	9.036	9.057	9.078	9.099	10	11	11	12
.96	9.120	9.141	9.162	9.183	9.204	9.226	9.247	9.268	9.290	9.311	12	13	13	14
.97	9.333	9.354	9.376	9.397	9.419	9.441	9.462	9.484	9.506	9.528	14	15	15	16
.98	9.550	9.572	9.594	9.616	9.638	9.661	9.683	9.705	9.727	9.750	16	17	18	18
.99	9.772	9.795	9.817	9.840	9.863	9.886	9.908	9.931	9.954	9.977	18	19	20	20

NOTE THE CHARACTERISTIC $\bar{1}$.

No.	0	1	2	3	4	5	6	7	8	9	INTERPOLATION TABLES.
1.00		$\bar{1}.9996$	$\bar{1}.9991$	$\bar{1}.9987$	$\bar{1}.9983$	$\bar{1}.9978$	$\bar{1}.9974$	$\bar{1}.9970$	$\bar{1}.9965$	$\bar{1}.9961$	
.01	$\bar{1}.9957$	9953	9948	9944	9940	9935	9931	9927	9923	9918	
.02	9914	9910	9905	9901	9897	9893	9887	9884	9880	9876	
.03	9872	9867	9863	9859	9855	9851	9846	9842	9838	9834	
.04	9830	9825	9821	9817	9813	9809	9805	9801	9796	9792	Use the first ten lines to avoid interpolating from 1.000 to 1.100.
1.05	$\bar{1}.9788$	$\bar{1}.9784$	$\bar{1}.9780$	$\bar{1}.9776$	$\bar{1}.9772$	$\bar{1}.9767$	$\bar{1}.9763$	$\bar{1}.9759$	$\bar{1}.9755$	$\bar{1}.9751$	
.06	9747	9743	9739	9735	9731	9727	9722	9718	9714	9710	
.07	9706	9702	9698	9694	9690	9686	9682	9678	9674	9670	
.08	9666	9662	9658	9654	9650	9646	9642	9638	9634	9630	
.09	9626	9622	9618	9614	9610	9606	9602	9598	9594	9590	
1.0	0.0000	$\bar{1}.9957$	$\bar{1}.9914$	$\bar{1}.9872$	$\bar{1}.9830$	$\bar{1}.9788$	$\bar{1}.9747$	$\bar{1}.9706$	$\bar{1}.9666$	$\bar{1}.9626$	**-44 -40 -36 -32**
.1	$\bar{1}.9586$	9547	9508	9469	9431	9393	9355	9318	9281	9245	4 4 4 3
.2	9208	9172	9136	9101	9066	9031	8996	8962	8928	8894	9 8 7 6
.3	8861	8827	8794	8761	8729	8697	8665	8633	8601	8570	13 12 11 10
.4	8539	8508	8477	8447	8416	8386	8356	8327	8297	8268	18 16 14 13
1.5	$\bar{1}.8239$	$\bar{1}.8210$	$\bar{1}.8182$	$\bar{1}.8153$	$\bar{1}.8125$	$\bar{1}.8097$	$\bar{1}.8069$	$\bar{1}.8041$	$\bar{1}.8013$	$\bar{1}.7986$	22 20 18 16
.6	7959	7932	7905	7878	7852	7825	7799	7773	7747	7721	26 24 22 19
.7	7696	7670	7645	7620	.7595	7570	7545	7520	7496	7471	31 28 25 22
.8	7447	7423	7399	7375	7352	7328	7305	7282	7258	7235	35 32 29 26
.9	7212	7190	7167	7144	7122	7100	7077	7055	7033	7011	40 36 32 29
2.0	$\bar{1}.6990$	$\bar{1}.6968$	$\bar{1}.6946$	$\bar{1}.6925$	$\bar{1}.6904$	$\bar{1}.6882$	$\bar{1}.6861$	$\bar{1}.6840$	$\bar{1}.6819$	$\bar{1}.6799$	**-28 -26 -24 -22**
.1	6778	6757	6737	6716	6696	6676	6655	6635	6615	6596	3 3 2 2
.2	6576	6556	6536	6517	6497	6478	6459	6440	6421	6402	6 5 5 4
.3	6383	6364	6345	6326	6308	6289	6271	6253	6234	6216	8 8 7 7
.4	6198	6180	6162	6144	6126	6108	6091	6073	6055	6038	11 10 10 9
2.5	$\bar{1}.6021$	$\bar{1}.6003$	$\bar{1}.5986$	$\bar{1}.5969$	$\bar{1}.5952$	$\bar{1}.5935$	$\bar{1}.5918$	$\bar{1}.5901$	$\bar{1}.5884$	$\bar{1}.5867$	14 13 12 11
.6	5850	5834	5817	5800	5784	5768	5751	5735	5719	5702	17 16 14 13
.7	5686	5670	5654	5638	5622	5607	5591	5575	5560	5544	20 18 17 15
.8	5528	5513	5498	5482	5467	5452	5436	5421	5406	5391	22 21 19 18
.9	5376	5361	5346	5331	5317	5302	5287	5272	5258	5243	25 23 22 20
3.0	$\bar{1}.5229$	$\bar{1}.5214$	$\bar{1}.5200$	$\bar{1}.5186$	$\bar{1}.5171$	$\bar{1}.5157$	$\bar{1}.5143$	$\bar{1}.5129$	$\bar{1}.5114$	$\bar{1}.5100$	**-18 -16 -14 -12**
.1	5086	5072	5058	5045	5031	5017	5003	4989	4976	4962	2 2 1 1
.2	4948	4935	4921	4908	4895	4881	4868	4855	4841	4828	4 3 3 2
.3	4815	4802	4789	4776	4763	4750	4737	4724	4711	4698	5 5 4 4
.4	4685	4672	4660	4647	4634	4622	4609	4597	4584	4572	7 6 6 5
3.5	$\bar{1}.4559$	$\bar{1}.4547$	$\bar{1}.4535$	$\bar{1}.4522$	$\bar{1}.4510$	$\bar{1}.4498$	$\bar{1}.4486$	$\bar{1}.4473$	$\bar{1}.4461$	$\bar{1}.4449$	9 8 7 6
.6	4437	4425	4413	4401	4389	4377	4365	4353	4342	4330	11 10 8 7
.7	4318	4306	4295	4283	4271	4260	4248	4237	4225	4214	13 11 10 8
.8	4202	4191	4179	4168	4157	4145	4134	4123	4112	4101	14 13 11 10
.9	4089	4078	4067	4056	4045	4034	4023	4012	4001	3990	16 14 13 11
4.0	$\bar{1}.3979$	$\bar{1}.3969$	$\bar{1}.3958$	$\bar{1}.3947$	$\bar{1}.3936$	$\bar{1}.3925$	$\bar{1}.3915$	$\bar{1}.3904$	$\bar{1}.3893$	$\bar{1}.3883$	**-11 -10 -9 -8**
.1	3872	3862	3851	3840	3830	3820	3809	3799	3788	3778	1 1 1 1
.2	3768	3757	3747	3737	3726	3716	3706	3696	3686	3675	2 2 2 2
.3	3665	3655	3645	3635	3625	3615	3605	3595	3585	3575	3 3 3 2
.4	3565	3556	3546	3536	3526	3516	3507	3497	3487	3478	4 4 4 3
4.5	$\bar{1}.3468$	$\bar{1}.3458$	$\bar{1}.3449$	$\bar{1}.3439$	$\bar{1}.3429$	$\bar{1}.3420$	$\bar{1}.3410$	$\bar{1}.3401$	$\bar{1}.3391$	$\bar{1}.3382$	6 5 5 4
.6	3372	3363	3354	3344	3335	3325	3316	3307	3298	3288	7 6 5 5
.7	3279	3270	3261	3251	3242	3233	3224	3215	3206	3197	8 7 6 6
.8	3188	3179	3170	3161	3152	3143	3134	3125	3116	3107	9 8 7 6
.9	3098	3089	3080	3071	3063	3054	3045	3036	3028	3019	10 9 8 7

FOUR PLACE COLOGARITHMS.

No.	0	1	2	3	4	5	6	7	8	9	Interpolation Tables.		
5.0	$\bar{1}$.3010	$\bar{1}$.3002	$\bar{1}$.2993	$\bar{1}$.2984	$\bar{1}$.2976	$\bar{1}$.2967	$\bar{1}$.2958	$\bar{1}$.2950	$\bar{1}$.2941	$\bar{1}$.2933	-9	-8	-7
.1	2924	2916	2907	2899	2890	2882	2874	2865	2857	2848	1	1	1
.2	2840	2832	2823	2815	2807	2798	2790	2782	2774	2765	2	2	1
.3	2757	2749	2741	2733	2725	2716	2708	2700	2692	2684	3	2	2
.4	2676	2668	2660	2652	2644	2636	2628	2620	2612	2604	4	3	3
5.5	$\bar{1}$.2596	$\bar{1}$.2588	$\bar{1}$.2581	$\bar{1}$.2573	$\bar{1}$.2565	$\bar{1}$.2557	$\bar{1}$.2549	$\bar{1}$.2541	$\bar{1}$.2534	$\bar{1}$.2526	5	4	4
.6	2518	2510	2503	2495	2487	2480	2472	2464	2457	2449	5	5	4
.7	2441	2434	2426	2418	2411	2403	2396	2388	2381	2373	6	6	5
.8	2366	2358	2351	2343	2336	2328	2321	2314	2306	2299	7	6	6
.9	2291	2284	2277	2269	2262	2255	2248	2240	2233	2226	8	7	6
6.0	$\bar{1}$.2218	$\bar{1}$.2211	$\bar{1}$.2204	$\bar{1}$.2197	$\bar{1}$.2190	$\bar{1}$.2182	$\bar{1}$.2175	$\bar{1}$.2168	$\bar{1}$.2161	$\bar{1}$.2154	-7		-6
.1	2147	2140	2132	2125	2118	2111	2104	2097	2090	2083	1		1
.2	2076	2069	2062	2055	2048	2041	2034	2027	2020	2013	1		1
.3	2007	2000	1993	1986	1979	1972	1965	1959	1952	1945	2		2
.4	1938	1931	1925	1918	1911	1904	1898	1891	1884	1878	3		2
6.5	$\bar{1}$.1871	$\bar{1}$.1864	$\bar{1}$.1858	$\bar{1}$.1851	$\bar{1}$.1844	$\bar{1}$.1838	$\bar{1}$.1831	$\bar{1}$.1824	$\bar{1}$.1818	$\bar{1}$.1811	4		3
.6	1805	1798	1791	1785	1778	1772	1765	1759	1752	1746	4		4
.7	1739	1733	1726	1720	1713	1707	1701	1694	1688	1681	5		4
.8	1675	1669	1662	1656	1649	1643	1637	1630	1624	1618	6		5
.9	1612	1605	1599	1593	1586	1580	1574	1568	1561	1555	6		5
7.0	$\bar{1}$.1549	$\bar{1}$.1543	$\bar{1}$.1537	$\bar{1}$.1530	$\bar{1}$.1524	$\bar{1}$.1518	$\bar{1}$.1512	$\bar{1}$.1506	$\bar{1}$.1500	$\bar{1}$.1494	-6		-5
.1	1487	1481	1475	1469	1463	1457	1451	1445	1439	1433	1		1
.2	1427	1421	1415	1409	1403	1397	1391	1385	1379	1373	1		1
.3	1367	1361	1355	1349	1343	1337	1331	1325	1319	1314	2		2
.4	1308	1302	1296	1290	1284	1278	1273	1267	1261	1255	2		2
7.5	$\bar{1}$.1249	$\bar{1}$.1244	$\bar{1}$.1238	$\bar{1}$.1232	$\bar{1}$.1226	$\bar{1}$.1221	$\bar{1}$.1215	$\bar{1}$.1209	$\bar{1}$.1203	$\bar{1}$.1198	3		3
.6	1192	1186	1180	1175	1169	1163	1158	1152	1146	1141	4		3
.7	1135	1129	1124	1118	1113	1107	1101	1096	1090	1085	4		4
.8	1079	1073	1068	1062	1057	1051	1046	1040	1035	1029	5		4
.9	1024	1018	1013	1007	1002	0996	0991	0985	0980	0975	5		5
8.0	$\bar{1}$.0969	$\bar{1}$.0964	$\bar{1}$.0958	$\bar{1}$.0953	$\bar{1}$.0947	$\bar{1}$.0942	$\bar{1}$.0937	$\bar{1}$.0931	$\bar{1}$.0926	$\bar{1}$.0921	-6		-5
.1	0915	0910	0904	0899	0894	0888	0883	0878	0872	0867	1		1
.2	0862	0857	0851	0846	0841	0835	0830	0825	0820	0814	1		1
.3	0809	0804	0799	0794	0788	0783	0778	0773	0768	0762	2		2
.4	0757	0752	0747	0742	0737	0731	0726	0721	0716	0711	2		2
8.5	$\bar{1}$.0706	$\bar{1}$.0701	$\bar{1}$.0696	$\bar{1}$.0691	$\bar{1}$.0685	$\bar{1}$.0680	$\bar{1}$.0675	$\bar{1}$.0670	$\bar{1}$.0665	$\bar{1}$.0660	3		3
.6	0655	0650	0645	0640	0635	0630	0625	0620	0615	0610	4		3
.7	0605	0600	0595	0590	0585	0580	0575	0570	0565	0560	4		4
.8	0555	0550	0545	0540	0535	0531	0526	0521	0516	0511	5		4
.9	0506	0501	0496	0491	0487	0482	0477	0472	0467	0462	5		5
9.0	$\bar{1}$.0458	$\bar{1}$.0453	$\bar{1}$.0448	$\bar{1}$.0443	$\bar{1}$.0438	$\bar{1}$.0434	$\bar{1}$.0429	$\bar{1}$.0424	$\bar{1}$.0419	$\bar{1}$.0414	-5		-4
.1	0410	0405	0400	0395	0391	0386	0381	0376	0372	0367	1		0
.2	0362	0357	0353	0348	0343	0339	0334	0329	0325	0320	1		1
.3	0315	0311	0306	0301	0297	0292	0287	0283	0278	0273	2		1
.4	0269	0264	0259	0255	0250	0246	0241	0237	0232	0227	2		2
9.5	$\bar{1}$.0223	$\bar{1}$.0218	$\bar{1}$.0214	$\bar{1}$.0209	$\bar{1}$.0205	$\bar{1}$.0200	$\bar{1}$.0195	$\bar{1}$.0191	$\bar{1}$.0186	$\bar{1}$.0182	3		2
.6	0177	0173	0168	0164	0159	0155	0150	0146	0141	0137	3		2
.7	0132	0128	0123	0119	0114	0110	0106	0101	0097	0092	4		3
.8	0088	0083	0079	0074	0070	0066	0061	0057	0052	0048	4		3
.9	0044	0039	0035	0031	0026	0022	0017	0013	0009	0004	5		4

TABLE OF FIVE PLACE LOGARITHMS,

CONTAINING

An Abbreviated Table for One and Two Place Numbers;
A Table for Five Place Numbers from 1.0 to 1.1 Avoiding Interpolation;
A Table for All Four Place Numbers with Interpolation Tables for the Fifth Place.

No.	log.	No.	log.	No.	log.	No.	log.	No.	log.
0	$-\infty$	**2.0**	.30 103	**4.0**	.60 206	**6.0**	.77 815	**8.0**	.90 309
1	.00 000	2.1	.32 222	4.1	.61 278	6.1	.78 533	8.1	.90 849
2	.30 103	2.2	.34 242	4.2	.62 325	6.2	.79 239	8.2	.91 381
3	.47 712	2.3	.36 173	4.3	.63 347	6.3	.79 934	8.3	.91 908
4	.60 206	2.4	.38 021	4.4	.64 345	6.4	.80 618	8.4	.92 428
5	.69 897	2.5	.39 794	4.5	.65 321	6.5	.81 291	8.5	.92 942
6	.77 815	2.6	.41 497	4.6	.66 276	6.6	.81 954	8.6	.93 450
7	.84 510	2.7	.43 136	4.7	.67 210	6.7	.82 607	8.7	.93 952
8	.90 309	2.8	.44 716	4.8	.68 124	6.8	.83 251	8.8	.94 448
9	.95 424	2.9	.46 240	4.9	.69 020	6.9	.83 885	8.9	.94 939
1.0	.00 000	**3.0**	.47 712	**5.0**	.69 897	**7.0**	.84 510	**9.0**	.95 424
1.1	.04 139	3.1	.49 136	5.1	.70 757	7.1	.85 126	9.1	.95 904
1.2	.07 918	3.2	.50 515	5.2	.71 600	7.2	.85 733	9.2	.96 379
1.3	.11 394	3.3	.51 851	5.3	.72 428	7.3	.86 332	9.3	.96 848
1.4	.14 613	3.4	.53 148	5.4	.73 239	7.4	.86 923	9.4	.97 313
1.5	.17 609	3.5	.54 407	5.5	.74 036	7.5	.87 506	9.5	.97 772
1.6	.20 412	3.6	.55 630	5.6	.74 819	7.6	.88 081	9.6	.98 227
1.7	.23 045	3.7	.56 820	5.7	.75 587	7.7	.88 649	9.7	.98 677
1.8	.25 527	3.8	.57 978	5.8	.76 343	7.8	.89 209	9.8	.99 123
1.9	.27 875	3.9	.59 106	5.9	.77 085	7.9	.89 763	9.9	.99 564

FIVE PLACE LOGARITHMS.

	No.	0	1	2	3	4	5	6	7	8	9
.00	**1.000**	.00 000	.00 004	.00 009	.00 013	.00 017	.00 022	.00 026	.00 030	.00 035	.00 039
.04	.001	00 043	00 048	00 052	00 056	00 061	00 065	00 069	00 074	00 078	00 082
	.002	00 087	00 091	00 095	00 100	00 104	00 108	00 113	00 117	00 121	00 126
	.003	00 130	00 134	00 139	00 143	00 147	00 152	00 156	00 160	00 165	00 169
	.004	00 173	00 178	00 182	00 186	00 191	00 195	00 199	00 204	00 208	00 212
	1.005	.00 217	.00 221	.00 225	.00 230	.00 234	.00 238	.00 243	.00 247	.00 251	.00 255
	.006	00 260	00 264	00 268	00 273	00 277	00 281	00 286	00 290	00 294	00 299
	.007	00 303	00 307	00 312	00 316	00 320	00 325	00 329	00 333	00 337	00 342
	.008	00 346	00 350	00 355	00 359	00 363	00 368	00 372	00 376	00 381	00 385
	.009	00 389	00 393	00 398	00 402	00 406	00 411	00 415	00 419	00 424	00 428
	1.010	.00 432	.00 436	.00 441	.00 445	.00 449	.00 454	.00 458	.00 462	.00 467	.00 471
	.011	00 475	00 479	00 484	00 488	00 492	00 497	00 501	00 505	00 509	00 514
	.012	00 518	00 522	00 527	00 531	00 535	00 540	00 544	00 548	00 552	00 557
	.013	00 561	00 565	00 570	00 574	00 578	00 582	00 587	00 591	00 595	00 600
	.014	00 604	00 608	00 612	00 617	00 621	00 625	00 629	00 634	00 638	00 642
	1.015	.00 647	.00 651	.00 655	.00 659	.00 664	.00 668	.00 672	.00 677	.00 681	.00 685
	.016	00 689	00 694	00 698	00 702	00 706	00 711	00 715	00 719	00 724	00 728
	.017	00 732	00 736	00 741	00 745	00 749	00 753	00 758	00 762	00 766	00 771
	.018	00 775	00 779	00 783	00 788	00 792	00 796	00 800	00 805	00 809	00 813
	.019	00 817	00 822	00 826	00 830	00 834	00 839	00 843	00 847	00 852	00 856
	1.020	.00 860	.00 864	.00 869	.00 873	.00 877	.00 881	.00 886	.00 890	.00 894	.00 898
	.021	00 903	00 907	00 911	00 915	00 920	00 924	00 928	00 932	00 937	00 941
	.022	00 945	00 949	00 954	00 958	00 962	00 966	00 971	00 975	00 979	00 983
	.023	00 988	00 992	00 996	01 000	01 005	01 009	01 013	01 017	01 022	01 026
	.024	01 030	01 034	01 038	01 043	01 047	01 051	01 055	01 060	01 064	01 068
	1.025	.01 072	.01 077	.01 081	.01 085	.01 089	.01 094	.01 098	.01 102	.01 106	.01 111
	.026	01 115	01 119	01 123	01 127	01 132	01 136	01 140	01 144	01 149	01 153
	.027	01 157	01 161	01 166	01 170	01 174	01 178	01 182	01 187	01 191	01 195
	.028	01 199	01 204	01 208	01 212	01 216	01 220	01 225	01 229	01 233	01 237
	.029	01 242	01 246	01 250	01 254	01 258	01 263	01 267	01 271	01 275	01 280
	1.030	.01 284	.01 288	.01 292	.01 296	.01 301	.01 305	.01 309	.01 313	.01 317	.01 322
	.031	01 326	01 330	01 334	01 338	01 343	01 347	01 351	01 355	01 360	01 364
	.032	01 368	01 372	01 376	01 381	01 385	01 389	01 393	01 397	01 402	01 406
	.033	01 410	01 414	01 418	01 423	01 427	01 431	01 435	01 439	01 444	01 448
	.034	01 452	01 456	01 460	01 465	01 469	01 473	01 477	01 481	01 486	01 490
	1.035	.01 494	.01 498	.01 502	.01 507	.01 511	.01 515	.01 519	.01 523	.01 528	.01 532
	.036	01 536	01 540	01 544	01 549	01 553	01 557	01 561	01 565	01 570	01 574
	.037	01 578	01 582	01 586	01 590	01 595	01 599	01 603	01 607	01 611	01 616
	.038	01 620	01 624	01 628	01 632	01 636	01 641	01 645	01 649	01 653	01 657
	.039	01 662	01 666	01 670	01 674	01 678	01 682	01 687	01 691	01 695	01 699
	1.040	.01 703	.01 708	.01 712	.01 716	.01 720	.01 724	.01 728	.01 733	.01 737	.01 741
	.041	01 745	01 749	01 753	01 758	01 762	01 766	01 770	01 774	01 778	01 783
	.042	01 787	01 791	01 795	01 799	01 803	01 808	01 812	01 816	01 820	01 824
	.043	01 828	01 833	01 837	01 841	01 845	01 849	01 853	01 858	01 862	01 866
	.044	01 870	01 874	01 878	01 883	01 887	01 891	01 895	01 899	01 903	01 907
	1.045	.01 912	.01 916	.01 920	.01 924	.01 928	.01 932	.01 937	.01 941	.01 945	.01 949
.00	.046	01 953	01 957	01 961	01 966	01 970	01 974	01 978	01 982	01 986	01 991
.04	.047	01 995	01 999	02 003	02 007	02 011	02 015	02 020	02 024	02 028	02 032
	.048	02 036	02 040	02 044	02 049	02 053	02 057	02 061	02 065	02 069	02 073
	.049	02 078	02 082	02 086	02 090	02 094	02 098	02 102	02 107	02 111	02 115

No.	0	1	2	3	4	5	6	7	8	9	
1.050	.02 119	.02 123	.02 127	.02 131	.02 135	.02 140	.02 144	.02 148	.02 152	.02 156	.00
.051	02 160	02 164	02 169	02 173	02 177	02 181	02 185	02 189	02 193	02 197	.04
.052	02 202	02 206	02 210	02 214	02 218	02 222	02 226	02 230	02 235	02 239	
.053	02 243	02 247	02 251	02 255	02 259	02 263	02 268	02 272	02 276	02 280	
.054	02 284	02 288	02 292	02 296	02 301	02 305	02 309	02 313	02 317	02 321	
1.055	.02 325	.02 329	.02 333	.02 338	.02 342	.02 346	.02 350	.02 354	.02 358	.02 362	
.056	02 366	02 371	02 375	02 379	02 383	02 387	02 391	02 395	02 399	02 403	
.057	02 407	02 412	02 416	02 420	02 424	02 428	02 432	02 436	02 440	02 444	
.058	02 449	02 453	02 457	02 461	02 465	02 469	02 473	02 477	02 481	02 485	
.059	02 490	02 494	02 498	02 502	02 506	02 510	02 514	02 518	02 522	02 526	
1.060	.02 531	.02 535	.02 539	.02 543	.02 547	.02 551	.02 555	.02 559	.02 563	.02 567	
.061	02 572	02 576	02 580	02 584	02 588	02 592	02 596	02 600	02 604	02 608	
.062	02 612	02 617	02 621	02 625	02 629	02 633	02 637	02 641	02 645	02 649	
.063	02 653	02 657	02 661	02 666	02 670	02 674	02 678	02 682	02 686	02 690	
.064	02 694	02 698	02 702	02 706	02 710	02 715	02 719	02 723	02 727	02 731	
1.065	.02 735	.02 739	.02 743	.02 747	.02 751	.02 755	.02 759	.02 763	.02 768	.02 772	
.066	02 776	02 780	02 784	02 788	02 792	02 796	02 800	02 804	02 808	02 812	
.067	02 816	02 821	02 825	02 829	02 833	02 837	02 841	02 845	02 849	02 853	
.068	02 857	02 861	02 865	02 869	02 873	02 877	02 882	02 886	02 890	02 894	
.069	02 898	02 902	02 906	02 910	02 914	02 918	02 922	02 926	02 930	02 934	
1.070	.02 938	.02 942	.02 946	.02 951	.02 955	.02 959	.02 963	.02 967	.02 971	.02 975	
.071	02 979	02 983	02 987	02 991	02 995	02 999	03 003	03 007	03 011	03 015	
.072	03 019	03 024	03 028	03 032	03 036	03 040	03 044	03 048	03 052	03 056	
.073	03 060	03 064	03 068	03 072	03 076	03 080	03 084	03 088	03 092	03 096	
.074	03 100	03 104	03 109	03 113	03 117	03 121	03 125	03 129	03 133	03 137	
1.075	.03 141	.03 145	.03 149	.03 153	.03 157	.03 161	.03 165	.03 169	.03 173	.03 177	
.076	03 181	03 185	03 189	03 193	03 197	03 201	03 205	03 209	03 214	03 218	
.077	03 222	03 226	03 230	03 234	03 238	03 242	03 246	03 250	03 254	03 258	
.078	03 262	03 266	03 270	03 274	03 278	03 282	03 286	03 290	03 294	03 298	
.079	03 302	03 306	03 310	03 314	03 318	03 322	03 326	03 330	03 334	03 338	
1.080	.03 342	.03 346	.03 350	.03 354	.03 358	.03 362	.03 366	.03 371	.03 375	.03 379	
.081	03 383	03 387	03 391	03 395	03 399	03 403	03 407	03 411	03 415	03 419	
.082	03 423	03 427	03 431	03 435	03 439	03 443	03 447	03 451	03 455	03 459	
.083	03 463	03 467	03 471	03 475	03 479	03 483	03 487	03 491	03 495	03 499	
.084	03 503	03 507	03 511	03 515	03 519	03 523	03 527	03 531	03 535	03 539	
1.085	.03 543	.03 547	.03 551	.03 555	.03 559	.03 563	.03 567	.03 571	.03 575	.03 579	
.086	03 583	03 587	03 591	03 595	03 599	03 603	03 607	03 611	03 615	03 619	
.087	03 623	03 627	03 631	03 635	03 639	03 643	03 647	03 651	03 655	03 659	
.088	03 663	03 667	03 671	03 675	03 679	03 683	03 687	03 691	03 695	03 699	
.089	03 703	03 707	03 711	03 715	03 719	03 723	03 727	03 731	03 735	03 739	
1.090	.03 743	.03 747	.03 751	.03 755	.03 759	.03 763	.03 767	.03 771	.03 775	.03 778	
.091	03 782	03 786	03 790	03 794	03 798	03 802	03 806	03 810	03 814	03 818	
.092	03 822	03 826	03 830	03 834	03 838	03 842	03 846	03 850	03 854	03 858	
.093	03 862	03 866	03 870	03 874	03 878	03 882	03 886	03 890	03 894	03 898	
.094	03 902	03 906	03 910	03 914	03 918	03 922	03 926	93 930	03 933	03 937	
1.095	.03 941	.03 945	.03 949	.03 953	.03 957	.03 961	.03 965	.03 969	.03 973	.03 977	
.096	03 981	03 985	03 989	03 993	03 997	04 001	04 005	04 009	04 013	04 017	.00
.097	04 021	04 025	04 029	04 033	04 036	04 040	04 044	04 048	04 052	04 056	.04
.098	04 060	04 064	04 068	04 072	04 076	04 080	04 084	04 088	04 092	04 096	
.099	04 100	04 104	04 108	04 112	04 116	04 120	04 123	04 127	04 131	04 135	

1.

FIVE PLACE LOGARITHMS.

	No.	0	1	2	3	4	5	6	7	8	9	INTERPO. TABLES.	
.00	**1.00**	**.00 000**	.00 043	.00 087	.00 130	.00 173	.00 217	.00 260	.00 303	.00 346	.00 389	For log 1.0 to log 1.1 Interpolated values are given on the preceding two pages.	
.30	.01	00 432	00 475	00 518	00 561	00 604	00 647	00 689	00 732	00 775	00 817		
	.02	00 860	00 903	00 945	00 988	**01 030**	01 072	01 115	01 157	01 199	01 242		
	.03	01 284	01 326	01 368	01 410	01 452	01 494	01 536	01 578	01 620	01 662		
	.04	01 703	01 745	01 787	01 828	01 870	01 912	01 953	01 995	**02 036**	02 078		
	1.05	.02 119	.02 160	.02 202	.02 243	.02 284	.02 325	.02 366	.02 407	.02 449	.02 490		
	.06	02 531	02 572	02 612	02 653	02 694	02 735	02 776	02 816	02 857	02 898		
	.07	02 938	02 979	**03 019**	03 060	03 100	03 141	03 181	03 222	03 262	03 302		
	.08	03 342	03 383	03 423	03 463	03 503	03 543	03 583	03 623	03 663	03 703		
	.09	03 743	03 782	03 822	03 862	03 902	03 941	03 981	**04 021**	04 060	04 100		
	1.10	.04 139	.04 179	.04 218	.04 258	.04 297	.04 336	.04 376	.04 415	.04 454	.04 493	**38**	**36**
	.11	04 532	04 571	04 610	04 650	04 689	04 727	04 766	04 805	04 844	04 883	4	4
	.12	04 922	04 961	04 999	**05 038**	05 077	05 115	05 154	05 192	05 231	05 269	8	7
	.13	05 308	05 346	05 385	05 423	05 461	05 500	05 538	05 576	05 614	05 652	11	11
	.14	05 690	05 729	05 767	05 805	05 843	05 881	05 918	05 956	05 994	**06 032**	15	14
	1.15	.06 070	.06 108	.06 145	.06 183	.06 221	.06 258	.06 296	.06 333	.06 371	.06 408	19	18
	.16	06 446	06 483	06 521	06 558	06 595	06 633	06 670	06 707	06 744	06 781	23	22
	.17	06 819	06 856	06 893	06 930	06 967	**07 004**	07 041	07 078	07 115	07 151	27	25
	.18	07 188	07 225	07 262	07 298	07 335	07 372	07 408	07 445	07 482	07 518	30	29
	.19	07 555	07 591	07 628	07 664	07 700	07 737	07 773	07 809	07 846	07 882	34	32
	1.20	.07 918	.07 954	.07 990	**.08 027**	.08 063	.08 099	.08 135	.08 171	.08 207	.08 243	**34**	**33**
	.21	08 279	08 314	08 350	08 386	08 422	08 458	08 493	08 529	08 565	08 600	3	3
	.22	08 636	08 672	08 707	08 743	08 778	08 814	08 849	08 884	08 920	08 955	7	7
	.23	08 991	**09 026**	09 061	09 096	09 132	09 167	09 202	09 237	09 272	09 307	10	10
	.24	09 342	09 377	09 412	09 447	09 482	09 517	09 552	09 587	09 621	09 656	14	13
	1.25	.09 691	.09 726	.09 760	.09 795	.09 830	.09 864	.09 899	.09 934	.09 968	**.10 003**	17	17
	.26	10 037	10 072	10 106	10 140	10 175	10 209	10 243	10 278	10 312	10 346	20	20
	.27	10 380	10 415	10 449	10 483	10 517	10 551	10 585	10 619	10 653	10 687	24	23
	.28	10 721	10 755	10 789	10 823	10 857	10 890	10 924	10 958	10 992	**11 025**	27	26
	.29	11 059	11 093	11 126	11 160	11 193	11 227	11 261	11 294	11 327	11 361	31	30
	1.30	.11 394	.11 428	.11 461	.11 494	.11 528	.11 561	.11 594	.11 628	.11 661	.11 694	**32**	**31**
	.31	11 727	11 760	11 793	11 826	11 860	11 893	11 926	11 959	11 992	**12 024**	3	3
	.32	12 057	12 090	12 123	12 156	12 189	12 222	12 254	12 287	12 320	12 352	6	6
	.33	12 385	12 418	12 450	12 483	12 516	12 548	12 581	12 613	12 646	12 678	10	9
	.34	12 710	12 743	12 775	12 808	12 840	12 872	12 905	12 937	12 969	**13 001**	13	12
	1.35	.13 033	.13 066	.13 098	.13 130	.13 162	.13 194	.13 226	.13 258	.13 290	.13 322	16	16
	.36	13 354	13 386	13 418	13 450	13 481	13 513	13 545	13 577	13 609	13 640	19	19
	.37	13 672	13 704	13 735	13 767	13 799	13 830	13 862	13 893	13 925	13 956	22	22
	.38	13 988	**14 019**	14 051	14 082	14 114	14 145	14 176	14 208	14 239	14 270	26	25
	.39	14 301	14 333	14 364	14 395	14 426	14 457	14 489	14 520	14 551	14 582	29	28
	1.40	.14 613	.14 644	.14 675	.14 706	.14 737	.14 768	.14 799	.14 829	.14 860	.14 891	**30**	**29**
	.41	14 922	14 953	14 983	**15 014**	15 045	15 076	15 106	15 137	15 168	15 198	3	3
	.42	15 229	15 259	15 290	15 320	15 351	15 381	15 412	15 442	15 473	15 503	6	6
	.43	15 534	15 564	15 594	15 625	15 655	15 685	15 715	15 746	15 776	15 806	9	9
	.44	15 836	15 866	15 897	15 927	15 957	15 987	**16 017**	16 047	16 077	16 107	12	12
	1.45	.16 137	.16 167	.16 197	.16 227	.16 256	.16 286	.16 316	.16 346	.16 376	.16 406	15	15
.00	.46	16 435	16 465	16 495	16 524	16 554	16 584	16 613	16 643	16 673	16 702	18	17
.30	.47	16 732	16 761	16 791	16 820	16 850	16 879	16 909	16 938	16 967	16 997	21	20
	.48	**17 026**	17 056	17 085	17 114	17 143	17 173	17 202	17 231	17 260	17 289	24	23
	.49	17 319	17 348	17 377	17 406	17 435	17 464	17 493	17 522	17 551	17 580	27	26

FIVE PLACE LOGARITHMS.

No.	0	1	2	3	4	5	6	7	8	9	INTERPO. TABLES.		
1.50	.17 609	.17 638	.17 667	.17 696	.17 725	.17 754	.17 782	.17 811	.17 840	.17 869	**29**	**27**	.00
.51	17 898	17 926	17 955	17 984	**18 013**	18 041	18 070	18 099	18 127	18 156	3	3	.30
.52	18 184	18 213	18 241	18 270	18 298	18 327	18 355	18 384	18 412	18 441	6	5	
.53	18 469	18 498	18 526	18 554	18 583	18 611	18 639	18 667	18 696	18 724	9	8	
.54	18 752	18 780	18 808	18 837	18 865	18 893	18 921	18 949	18 977	**19 005**	12	11	
1.55	.19 033	.19 061	.19 089	.19 117	.19 145	.19 173	.19 201	.19 229	.19 257	.19 285	15	14	
.56	19 312	19 340	19 368	19 396	19 424	19 451	19 479	19 507	19 535	19 562	17	16	
.57	19 590	19 618	19 645	19 673	19 700	19 728	19 756	19 783	19 811	19 838	20	19	
.58	19 866	19 893	19 921	19 948	19 976	**20 003**	20 030	20 058	20 085	20 112	23	22	
.59	20 140	20 167	20 194	20 222	20 249	20 276	20 303	20 330	20 358	20 385	26	24	
1.60	.20 412	.20 439	.20 466	.20 493	.20 520	.20 548	.20 575	.20 602	.20 629	.20 656	**26**	**25**	
.61	20 683	20 710	20 737	20 763	20 790	20 817	20 844	20 871	20 898	20 925	3	3	
.62	20 952	20 978	**21 005**	21 032	21 059	21 085	21 112	21 139	21 165	21 192	5	5	
.63	21 219	21 245	21 272	21 299	21 325	21 352	21 378	21 405	21 431	21 458	8	8	
.64	21 484	21 511	21 537	21 564	21 590	21 617	21 643	21 669	21 696	21 722	10	10	
1.65	.21 748	.21 775	.21 801	.21 827	.21 854	.21 880	.21 906	.21 932	.21 958	.21 985	13	13	
.66	**22 011**	22 037	22 063	22 089	22 115	22 141	22 167	22 194	22 220	22 246	16	15	
.67	22 272	22 298	22 324	22 350	22 376	22 401	22 427	22 453	22 479	22 505	18	18	
.68	22 531	22 557	22 583	22 608	22 634	22 660	22 686	22 712	22 737	22 763	21	20	
.69	22 789	22 814	22 840	22 866	22 891	22 917	22 943	22 968	22 994	**23 019**	23	23	
1.70	.23 045	.23 070	.23 096	.23 121	.23 147	.23 172	.23 198	.23 223	.23 249	.23 274	**25**	**24**	
.71	23 300	23 325	23 350	23 376	23 401	23 426	23 452	23 477	23 502	23 528	3	2	
.72	23 553	23 578	23 603	23 629	23 654	23 679	23 704	23 729	23 754	23 779	5	5	
.73	23 805	23 830	23 855	23 880	23 905	23 930	23 955	23 980	**24 005**	24 030	8	7	
.74	24 055	24 080	24 105	24 130	24 155	24 180	24 204	24 229	24 254	24 279	10	10	
1.75	.24 304	.24 329	.24 353	.24 378	.24 403	.24 428	.24 452	.24 477	.24 502	.24 527	13	12	
.76	24 551	24 576	24 601	24 625	24 650	24 674	24 699	24 724	24 748	24 773	15	14	
.77	24 797	24 822	24 846	24 871	24 895	24 920	24 944	24 969	24 993	**25 018**	18	17	
.78	25 042	25 066	25 091	25 115	25 139	25 164	25 188	25 212	25 237	25 261	20	19	
.79	25 285	25 310	25 334	25 358	25 382	25 406	25 431	25 455	25 479	25 503	23	22	
1.80	.25 527	.25 551	.25 575	.25 600	.25 624	.25 648	.25 672	.25 696	.25 720	.25 744	**24**	**23**	
.81	25 768	25 792	25 816	25 840	25 864	25 888	25 912	25 935	25 959	25 983	2	2	
.82	**26 007**	26 031	26 055	26 079	26 102	26 126	26 150	26 174	26 198	26 221	5	5	
.83	26 245	26 269	26 293	26 316	26 340	26 364	26 387	26 411	26 435	26 458	7	7	
.84	26 482	26 505	26 529	26 553	26 576	26 600	26 623	26 647	26 670	26 694	10	9	
1.85	.26 717	.26 741	.26 764	.26 788	.26 811	.26 834	.26 858	.26 881	.26 905	.26 928	12	12	
.86	26 951	26 975	26 998	**27 021**	27 045	27 068	27 091	27 114	27 138	27 161	14	14	
.87	27 184	27 207	27 231	27 254	27 277	27 300	27 323	27 346	27 370	27 393	17	16	
.88	27 416	27 439	27 462	27 485	27 508	27 531	27 554	27 577	27 600	27 623	19	18	
.89	27 646	27 669	27 692	27 715	27 738	27 761	27 784	27 807	27 830	27 852	22	20	
1.90	.27 875	.27 898	.27 921	.27 944	.27 967	.27 989	**.28 012**	.28 035	.28 058	.28 081	**22**	**21**	
.91	28 103	28 126	28 149	28 171	28 194	28 217	28 240	28 262	28 285	28 307	2	2	
.92	28 330	28 353	28 375	28 398	28 421	28 443	28 466	28 488	28 511	28 533	4	4	
.93	28 556	28 578	28 601	28 623	28 646	28 668	28 691	28 713	28 735	28 758	7	6	
.94	28 780	28 803	28 825	28 847	28 870	28 892	28 914	28 937	28 959	28 981	9	8	
.95	**.29 003**	.29 026	.29 048	.29 070	.29 092	.29 115	.29 137	.29 159	.29 181	.29 203	11	11	
.96	29 226	29 248	29 270	29 292	29 314	29 336	29 358	29 380	29 403	29 425	13	13	.00
.97	29 447	29 469	29 491	29 513	29 535	29 557	29 579	29 601	29 623	29 645	15	15	.30
.98	29 667	29 688	29 710	29 732	29 754	29 776	29 798	29 820	29 842	29 863	18	17	
.99	29 885	29 907	29 929	29 951	29 973	29 994	**30 016**	30 038	30 060	30 081	20	19	

FIVE PLACE LOGARITHMS.

	No.	0	1	2	3	4	5	6	7	8	9	INTP. TAB.
.30	**2.00**	.30 103	.30 125	.30 146	.30 168	.30 190	.30 211	.30 233	.30 255	.30 276	.30 298	**21**
.47	.01	30 320	30 341	30 363	30 384	30 406	30 428	30 449	30 471	30 492	30 514	2
	.02	30 535	30 557	30 578	30 600	30 621	30 643	30 664	30 685	30 707	30 728	4
	.03	30 750	30 771	30 792	30 814	30 835	30 856	30 878	30 899	30 920	30 942	6
	.04	30 963	30 984	**31 006**	31 027	31 048	31 069	31 091	31 112	31 133	31 154	8
	2.05	.31 175	.31 197	.31 218	.31 239	.31 260	.31 281	.31 302	.31 323	.31 345	.31 366	11
	.06	31 387	31 408	31 429	31 450	31 471	31 492	31 513	31 534	31 555	31 576	13
	.07	31 597	31 618	31 639	31 660	31 681	31 702	31 723	31 744	31 765	31 785	15
	.08	31 806	31 827	31 848	31 869	31 890	31 911	31 931	31 952	31 973	31 994	17
	.09	**32 015**	32 035	32 056	32 077	32 098	32 118	32 139	32 160	32 181	32 201	19
	2.10	.32 222	.32 243	.32 263	.32 284	.32 305	.32 325	.32 346	.32 366	.32 387	.32 408	**20**
	.11	32 428	32 449	32 469	32 490	32 510	32 531	32 552	32 572	32 593	32 613	2
	.12	32 634	32 654	32 675	32 695	32 715	32 736	32 756	32 777	32 797	32 818	4
	.13	32 838	32 858	32 879	32 899	32 919	32 940	32 960	32 980	**33 001**	33 021	6
	.14	33 041	33 062	33 082	33 102	33 122	33 143	33 163	33 183	33 203	33 224	8
	2.15	.33 244	.33 264	.33 284	.33 304	.33 325	.33 345	.33 365	.33 385	.33 405	.33 425	10
	.16	33 445	33 465	33 486	33 506	33 526	33 546	33 566	33 586	33 606	33 626	12
	.17	33 646	33 666	33 686	33 706	33 726	33 746	33 766	33 786	33 806	33 826	14
	.18	33 846	33 866	33 885	33 905	33 925	33 945	33 965	33 985	**34 005**	34 025	16
	.19	34 044	34 064	34 084	34 104	34 124	34 143	34 163	34 183	34 203	34 223	18
	2.20	.34 242	.34 262	.34 282	.34 301	.34 321	.34 341	.34 361	.34 380	.34 400	.34 420	**19**
	.21	34 439	34 459	34 479	34 498	34 518	34 537	34 557	34 577	34 596	34 616	2
	.22	34 635	34 655	34 674	34 694	34 713	34 733	34 753	34 772	34 792	34 811	4
	.23	34 830	34 850	34 869	34 889	34 908	34 928	34 947	34 967	34 986	**35 005**	6
	.24	35 025	35 044	35 064	35 083	35 102	35 122	35 141	35 160	35 180	35 199	8
	2.25	.35 218	.35 238	.35 257	.35 276	.35 295	.35 315	.35 334	.35 353	.35 372	.35 392	10
	.26	35 411	35 430	35 449	35 468	35 488	35 507	35 526	35 545	35 564	35 583	11
	.27	35 603	35 622	35 641	35 660	35 679	35 698	35 717	35 736	35 755	35 774	13
	.28	35 793	35 813	35 832	35 851	35 870	35 889	35 908	35 927	35 946	35 965	15
	.29	35 984	**36 003**	36 021	36 040	36 059	36 078	36 097	36 116	36 135	36 154	17
	2.30	.36 173	.36 192	.36 211	.36 229	.36 248	.36 267	.36 286	.36 305	.36 324	.36 342	**18**
	.31	36 361	36 380	36 399	36 418	36 436	36 455	36 474	36 493	36 511	36 530	2
	.32	36 549	36 568	36 586	36 605	36 624	36 642	36 661	36 680	36 698	36 717	4
	.33	36 736	36 754	36 773	36 791	36 810	36 829	36 847	36 866	36 884	36 903	5
	.34	36 922	36 940	36 959	36 977	36 996	**07 014**	37 033	37 051	37 070	37 088	7
	2.35	.37 107	.37 125	.37 144	.37 162	.37 181	.37 199	.37 218	.37 236	.37 254	.37 273	9
	.36	37 291	37 310	37 328	37 346	37 365	37 383	37 401	37 420	37 438	37 457	11
	.37	37 475	37 493	37 511	37 530	37 548	37 566	37 585	37 603	37 621	37 639	13
	.38	37 658	37 676	37 694	37 712	37 731	37 749	37 767	37 785	37 803	37 822	14
	.39	37 840	37 858	37 876	37 894	37 912	37 931	37 949	37 967	37 985	**38 003**	16
	2.40	.38 021	.38 039	.38 057	.38 075	.38 093	.38 112	.38 130	.38 148	.38 166	.38 184	**17**
	.41	38 202	38 220	38 238	38 256	38 274	38 292	38 310	38 328	38 346	38 364	2
	.42	38 382	38 399	38 417	38 435	38 453	38 471	38 489	38 507	38 525	38 543	3
	.43	38 561	38 578	38 596	38 614	38 632	38 650	38 668	38 686	38 703	38 721	5
	.44	38 739	38 757	38 775	38 792	38 810	38 828	38 846	38 863	38 881	38 899	7
	2.45	.38 917	.38 934	.38 952	.38 970	.38 987	**.39 005**	.39 023	.39 041	.39 058	.39 076	9
.30	.46	39 094	39 111	39 129	39 146	39 164	39 182	39 199	39 217	39 235	39 252	10
.47	.47	39 270	39 287	39 305	39 322	39 340	39 358	39 375	39 393	39 410	39 428	12
	.48	39 445	39 463	39 480	39 498	39 515	39 533	39 550	39 568	39 585	39 602	14
	.49	39 620	39 637	39 655	39 672	39 690	39 707	39 724	39 742	39 759	39 777	15

No.	0	1	2	3	4	5	6	7	8	9	INTP. TAB.	
2.50	.39 794	.39 811	.39 829	.39 846	.39 863	.39 881	.39 898	.39 915	.39 933	.39 950	**18**	**.30**
.51	39 967	39 985	**40 002**	40 019	40 037	40 054	40 071	40 088	40 106	40 123	2	**.47**
.52	40 140	40 157	40 175	40 192	40 209	40 226	40 243	40 261	40 278	40 295	4	
.53	40 312	40 329	40 346	40 364	40 381	40 398	40 415	40 432	40 449	40 466	5	
.54	40 483	40 500	40 518	40 535	40 552	40 569	40 586	40 603	40 620	40 637	7	
2.55	.40 654	.40 671	.40 688	.40 705	.40 722	.40 739	.40 756	.40 773	.40 790	.40 807	9	
.56	40 824	40 841	40 858	40 875	40 892	40 909	40 926	40 943	40 960	40 976	11	
.57	40 993	**41 010**	41 027	41 044	41 061	41 078	41 095	41 111	41 128	41 145	13	
.58	41 162	41 179	41 196	41 212	41 229	41 246	41 263	41 280	41 296	41 313	14	
.59	41 330	41 347	41 363	41 380	41 397	41 414	41 430	41 447	41 464	41 481	16	
2.60	.41 497	.41 514	.41 531	.41 547	.41 564	.41 581	.41 597	.41 614	.41 631	.41 647	**17**	
.61	41 664	41 681	41 697	41 714	41 731	41 747	41 764	41 780	41 797	41 814	2	
.62	41 830	41 847	41 863	41 880	41 896	41 913	41 929	41 946	41 963	41 979	3	
.63	41 996	**42 012**	42 029	42 045	42 062	42 078	42 095	42 111	42 127	42 144	5	
.64	42 160	42 177	42 193	42 210	42 226	42 243	42 259	42 275	42 292	42 308	7	
2.65	.42 325	.42 341	.42 357	.42 374	.42 390	.42 406	.42 423	.42 439	.42 455	.42 472	9	
.66	42 488	42 504	42 521	42 537	42 553	42 570	42 586	42 602	42 619	42 635	10	
.67	42 651	42 667	42 684	42 700	42 716	42 732	42 749	42 765	42 781	42 797	12	
.68	42 813	42 830	42 846	42 862	42 878	42 894	42 911	42 927	42 943	42 959	14	
.69	42 975	42 991	**43 008**	43 024	43 040	43 056	43 072	43 088	43 104	43 120	15	
2.70	.43 136	.43 152	.43 169	.43 185	.43 201	.43 217	.43 233	.43 249	.43 265	.43 281	**16**	
.71	43 297	43 313	43 329	43 345	43 361	43 377	43 393	43 409	43 425	43 441	2	
.72	43 457	43 473	43 489	43 505	43 521	43 537	43 553	43 569	43 584	43 600	3	
.73	43 616	43 632	43 648	43 664	43 680	43 696	43 712	43 727	43 743	43 759	5	
.74	43 775	43 791	43 807	43 823	43 838	43 854	43 870	43 886	43 902	43 917	6	
2.75	.43 933	.43 949	.43 965	.43 981	.43 996	**.44 012**	.44 028	.44 044	.44 059	.44 075	8	
.76	44 091	44 107	44 122	44 138	44 154	44 170	44 185	44 201	44 217	44 232	10	
.77	44 248	44 264	44 279	44 295	44 311	44 326	44 342	44 358	44 373	44 389	11	
.78	44 404	44 420	44 436	44 451	44 467	44 483	44 498	44 514	44 529	44 545	13	
.79	44 560	44 576	44 592	44 607	44 623	44 638	44 654	44 669	44 685	44 700	14	
2.80	.44 716	.44 731	.44 747	.44 762	.44 778	.44 793	.44 809	.44 824	.44 840	.44 855	**15**	
.81	44 871	44 886	44 902	44 917	44 932	44 948	44 963	44 979	44 994	**45 010**	2	
.82	45 025	45 040	45 056	45 071	45 086	45 102	45 117	45 133	45 148	45 163	3	
.83	45 179	45 194	45 209	45 225	45 240	45 255	45 271	45 286	45 301	45 317	5	
.84	45 332	45 347	45 362	45 378	45 393	45 408	45 423	45 439	45 454	45 469	6	
2.85	.45 484	.45 500	.45 515	.45 530	.45 545	.45 561	.45 576	.45 591	.45 606	.45 621	8	
.86	45 637	45 652	45 667	45 682	45 697	45 712	45 728	45 743	45 758	45 773	9	
.87	45 788	45 803	45 818	45 834	45 849	45 864	45 879	45 894	45 909	45 924	11	
.88	45 939	45 954	45 969	45 984	**46 000**	46 015	46 030	46 045	46 060	46 075	12	
.89	46 090	46 105	46 120	46 135	46 150	46 165	46 180	46 195	46 210	46 225	14	
2.90	.46 240	.46 255	.46 270	.46 285	.46 300	.46 315	.46 330	.46 345	.46 359	.46 374	**14**	
.91	46 389	46 404	46 419	46 434	46 449	46 464	46 479	46 494	46 509	46 523	1	
.92	46 538	46 553	46 568	46 583	46 598	46 613	46 627	46 642	46 657	46 672	3	
.93	46 687	46 702	46 716	46 731	46 746	46 761	46 776	46 790	46 805	46 820	4	
.94	46 835	46 850	46 864	46 879	46 894	46 909	46 923	46 938	46 953	46 967	6	
2.95	.46 982	.46 997	**.47 012**	.47 026	.47 041	.47 056	.47 070	.47 085	.47 100	.47 114	7	
.96	47 129	47 144	47 159	47 173	47 188	47 202	47 217	47 232	47 246	47 261	8	**.30**
.97	47 276	47 290	47 305	47 319	47 334	47 349	47 363	47 378	47 392	47 407	10	**.47**
.98	47 422	47 436	47 451	47 465	47 480	47 494	47 509	47 524	47 538	47 553	11	
.99	47 567	47 582	47 596	47 611	47 625	47 640	47 654	47 669	47 683	47 698	13	

FIVE PLACE LOGARITHMS.

	No.	0	1	2	3	4	5	6	7	8	9	INTP. TAB.
.47	**3.00**	.47 712	.47 727	.47 741	.47 756	.47 770	.47 784	.47 799	.47 813	.47 828	.47 842	**15**
.60	.01	47 857	47 871	47 885	47 900	47 914	47 929	47 943	47 958	47 972	47 986	2
	.02	**48 001**	48 015	48 029	48 044	48 058	48 073	48 087	48 101	48 116	48 130	3
	.03	48 144	48 159	48 173	48 187	48 202	48 216	48 230	48 244	48 259	48 273	5
	.04	48 287	48 302	48 316	48 330	48 344	48 359	48 373	48 387	48 401	48 416	6
	3.05	.48 430	.48 444	.48 458	.48 473	.48 487	.48 501	.48 515	.48 530	.48 544	.48 558	8
	.06	48 572	48 586	48 601	48 615	48 629	48 643	48 657	48 671	48 686	48 700	9
	.07	48 714	48 728	48 742	48 756	48 770	48 785	48 799	48 813	48 827	48 841	11
	.08	48 855	48 869	48 883	48 897	48 911	48 926	48 940	48 954	48 968	48 982	12
	.09	48 996	**49 010**	49 024	49 038	49 052	49 066	49 080	49 094	49 108	49 122	14
	3.10	.49 136	.49 150	.49 164	.49 178	.49 192	.49 206	.49 220	.49 234	.49 248	.49 262	**14**
	.11	49 276	49 290	49 304	49 318	49 332	49 346	49 360	49 374	49 388	49 402	1
	.12	49 415	49 429	49 443	49 457	49 471	49 485	49 499	49 513	49 527	49 541	3
	.13	49 554	49 568	49 582	49 596	49 610	49 624	49 638	49 651	49 665	49 679	4
	.14	49 693	49 707	49 721	49 734	49 748	49 762	49 776	49 790	49 803	49 817	6
	3.15	.49 831	.49 845	.49 859	.49 872	.49 886	.49 900	.49 914	.49 927	.49 941	.49 955	7
	.16	49 969	49 982	49 996	**50 010**	50 024	50 037	50 051	50 065	50 079	50 092	8
	.17	50 106	50 120	50 133	50 147	50 161	50 174	50 188	50 202	50 215	50 229	10
	.18	50 243	50 256	50 270	50 284	50 297	50 311	50 325	50 338	50 352	50 365	11
	.19	50 379	50 393	50 406	50 420	50 433	50 447	50 461	50 474	50 488	50 501	13
	3.20	.50 515	.50 529	.50 542	.50 556	.50 569	.50 583	.50 596	.50 610	.50 623	.50 637	**13**
	.21	50 651	50 664	50 678	50 691	50 705	50 718	50 732	50 745	50 759	50 772	1
	.22	50 786	50 799	50 813	50 826	50 840	50 853	50 866	50 880	50 893	50 907	3
	.23	50 920	50 934	50 947	50 961	50 974	50 987	**51 001**	51 014	51 028	51 041	4
	.24	51 055	51 068	51 081	51 095	51 108	51 121	51 135	51 148	51 162	51 175	5
	3.25	.51 188	.51 202	.51 215	.51 228	.51 242	.51 255	.51 268	.51 282	.51 295	.51 308	7
	.26	51 322	51 335	51 348	51 362	51 375	51 388	51 402	51 415	51 428	51 441	8
	.27	51 455	51 468	51 481	51 495	51 508	51 521	51 534	51 548	51 561	51 574	9
	.28	51 587	51 601	51 614	51 627	51 640	51 654	51 667	51 680	51 693	51 706	10
	.29	51 720	51 733	51 746	51 759	51 772	51 786	51 799	51 812	51 825	51 838	12
	3.30	.51 851	.51 865	.51 878	.51 891	.51 904	.51 917	.51 930	.51 943	.51 957	.51 970	**13**
	.31	51 983	51 996	**52 009**	52 022	52 035	52 048	52 061	52 075	52 088	52 101	1
	.32	52 114	52 127	52 140	52 153	52 166	52 179	52 192	52 205	52 218	52 231	3
	.33	52 244	52 257	52 270	52 284	52 297	52 310	52 323	52 336	52 349	52 362	4
	.34	52 375	52 388	52 401	52 414	52 427	52 440	52 453	52 466	52 479	52 492	5
	3.35	.52 504	.52 517	.52 530	.52 543	.52 556	.52 569	.52 582	.52 595	.52 608	.52 621	7
	.36	52 634	52 647	52 660	52 673	52 686	52 699	52 711	52 724	52 737	52 750	8
	.37	52 763	52 776	52 789	52 802	52 815	52 827	52 840	52 853	52 866	52 879	9
	.38	52 892	52 905	52 917	52 930	52 943	52 956	52 969	52 982	52 994	**53 007**	10
	.39	53 020	53 033	53 046	53 058	53 071	53 084	53 097	53 110	53 122	53 135	12
	3.40	.53 148	.53 161	.53 173	.53 186	.53 199	.53 212	.53 224	.53 237	.53 250	.53 263	**12**
	.41	53 275	53 288	53 301	53 314	53 326	53 339	53 352	53 364	53 377	53 390	1
	.42	53 403	53 415	53 428	53 441	53 453	53 466	53 479	53 491	53 504	53 517	2
	.43	53 529	53 542	53 555	53 567	53 580	53 593	53 605	53 618	53 631	53 643	4
	.44	53 656	53 668	53 681	53 694	53 706	53 719	53 732	53 744	53 757	53 769	5
	3.45	.53 782	.53 794	.53 807	.53 820	.53 832	.53 845	.53 857	.53 870	.53 882	.53 895	6
.47	.46	53 908	53 920	53 933	53 945	53 958	53 970	53 983	53 995	**54 008**	54 020	7
.60	.47	54 033	54 045	54 058	54 070	54 083	54 095	54 108	54 120	54 133	54 145	8
	.48	54 158	54 170	54 183	54 195	54 208	54 220	54 233	54 245	54 258	54 270	10
	.49	54 283	54 295	54 307	54 320	54 332	54 345	54 357	54 370	54 382	54 394	11

No.	0	1	2	3	4	5	6	7	8	9	INTP. TAB.	
3.50	.54 407	.54 419	.54 432	.54 444	.54 456	.54 469	.54 481	.54 494	.54 506	.54 518	**13**	.47
.51	54 531	54 543	54 555	54 568	54 580	54 593	54 605	54 617	54 630	54 642	1	.60
.52	54 654	54 667	54 679	54 691	54 704	54 716	54 728	54 741	54 753	54 765	3	
.53	54 777	54 790	54 802	54 814	54 827	54 839	54 851	54 864	54 876	54 888	4	
.54	54 900	54 913	54 925	54 937	54 949	54 962	54 974	54 986	54 998	**55 011**	5	
3.55	.55 023	.55 035	.55 047	.55 060	.55 072	.55 084	.55 096	.55 108	.55 121	.55 133	7	
.56	55 145	55 157	55 169	55 182	55 194	55 206	55 218	55 230	55 242	55 255	8	
.57	55 267	55 279	55 291	55 303	55 315	55 328	55 340	55 352	55 364	55 376	9	
.58	55 388	55 400	55 413	55 425	55 437	55 449	55 461	55 473	55 485	55 497	10	
.59	55 509	55 522	55 534	55 546	55 558	55 570	55 582	55 594	55 606	55 618	12	
3.60	.55 630	.55 642	.55 654	.55 666	.55 678	.55 691	.55 703	.55 715	.55 727	.55 739	**12**	
.61	55 751	55 763	55 775	55 787	55 799	55 811	55 823	55 835	55 847	55 859	1	
.62	55 871	55 883	55 895	55 907	55 919	55 931	55 943	55 955	55 967	55 979	2	
.63	55 991	**56 003**	56 015	56 027	56 038	56 050	56 062	56 074	56 086	56 098	4	
.64	56 110	56 122	56 134	56 146	56 158	56 170	56 182	56 194	56 205	56 217	5	
3.65	.56 229	.56 241	.56 253	.56 265	.56 277	.56 289	.56 301	.56 312	.56 324	.56 336	6	
.66	56 348	56 360	56 372	56 384	56 396	56 407	56 419	56 431	56 443	56 455	7	
.67	56 467	56 478	56 490	56 502	56 514	56 526	56 538	56 549	56 561	56 573	8	
.68	56 585	56 597	56 608	56 620	56 632	56 644	56 656	56 667	56 679	56 691	10	
.69	56 703	56 714	56 726	56 738	56 750	56 761	56 773	56 785	56 797	56 808	11	
3.70	.56 820	.56 832	.56 844	.56 855	.56 867	.56 879	.56 891	.56 902	.56 914	.56 926	**12**	
.71	56 937	56 949	56 961	56 972	56 984	56 996	**57 008**	57 019	57 031	57 043	1	
.72	57 054	57 066	57 078	57 089	57 101	57 113	57 124	57 136	57 148	57 159	2	
.73	57 171	57 183	57 194	57 206	57 217	57 229	57 241	57 252	57 264	57 276	4	
.74	57 287	57 299	57 310	57 322	57 334	57 345	57 357	57 368	57 380	57 392	5	
3.75	.57 403	.57 415	.57 426	.57 438	.57 449	.57 461	.57 473	.57 484	.57 496	.57 507	6	
.76	57 519	57 530	57 542	57 553	57 565	57 576	57 588	57 600	57 611	57 623	7	
.77	57 634	57 646	57 657	57 669	57 680	57 692	57 703	57 715	57 726	57 738	8	
.78	57 749	57 761	57 772	57 784	57 795	57 807	57 818	57 830	57 841	57 852	10	
.79	57 864	57 875	57 887	57 898	57 910	57 921	57 933	57 944	57 955	57 967	11	
3.80	.57 978	.57 990	**.58 001**	.58 013	.58 024	.58 035	.58 047	.58 058	.58 070	.58 081	**11**	
.81	58 092	58 104	58 115	58 127	58 138	58 149	58 161	58 172	58 184	58 195	1	
.82	58 206	58 218	58 229	58 240	58 252	58 263	58 274	58 286	58 297	58 309	2	
.83	58 320	58 331	58 343	58 354	58 365	58 377	58 388	58 399	58 410	58 422	3	
.84	58 433	58 444	58 456	58 467	58 478	58 490	58 501	58 512	58 524	58 535	4	
3.85	.58 546	.58 557	.58 569	.58 580	.58 591	.58 602	.58 614	.58 625	.58 636	.58 647	6	
.86	58 659	58 670	58 681	58 692	58 704	58 715	58 726	58 737	58 749	58 760	7	
.87	58 771	58 782	58 794	58 805	58 816	58 827	58 838	58 850	58 861	58 872	8	
.88	58 883	58 894	58 906	58 917	58 928	58 939	58 950	58 961	58 973	58 984	9	
.89	58 995	**59 006**	59 017	59 028	59 040	59 051	59 062	59 073	59 084	59 095	10	
3.90	.59 106	.59 118	.59 129	.59 140	.59 151	.59 162	.59 173	.59 184	.59 195	.59 207	**11**	
.91	59 218	59 229	59 240	59 251	59 262	59 273	59 284	59 295	59 306	59 318	1	
.92	59 329	59 340	59 351	59 362	59 373	59 384	59 395	59 406	59 417	59 428	2	
.93	59 439	59 450	59 461	59 472	59 483	59 494	59 506	59 517	59 528	59 539	3	
.94	59 550	59 561	59 572	59 583	59 594	59 605	59 616	59 627	59 638	59 649	4	
3.95	.59 660	.59 671	.59 682	.59 693	.59 704	.59 715	.59 726	.59 737	.59 748	.59 759	6	
.96	59 770	59 780	59 791	59 802	59 813	59 824	59 835	59 846	59 857	59 868	7	.47
.97	59 879	59 890	59 901	59 912	59 923	59 934	59 945	59 956	59 966	59 977	8	.60
.98	59 988	59 999	**60 010**	60 021	60 032	60 043	60 054	60 065	60 076	60 086	9	
.99	60 097	60 108	60 119	60 130	60 141	60 152	60 163	60 173	60 184	60 195	10	

	No.	0	1	2	3	4	5	6	7	8	9	INTP. TAB.
.60	**4.00**	.60 206	.60 217	.60 228	.60 239	.60 249	.60 260	.60 271	.60 282	.60 293	.60 304	**11**
.69	.01	60 314	60 325	60 336	60 347	60 358	60 369	60 379	60 390	60 401	60 412	1
	.02	60 423	60 433	60 444	60 455	60 466	60 477	60 487	60 498	60 509	60 520	2
	.03	60 531	60 541	60 552	60 563	60 574	60 584	60 595	60 606	60 617	60 627	3
	.04	60 638	60 649	60 660	60 670	60 681	60 692	60 703	60 713	60 724	60 735	4
	4.05	.60 746	.60 756	.60 767	.60 778	.60 788	.60 799	.60 810	.60 821	.60 831	.60 842	6
	.06	60 853	60 863	60 874	60 885	60 895	60 906	60 917	60 927	60 938	60 949	7
	.07	60 959	60 970	60 981	60 991	**61 002**	61 013	61 023	61 034	61 045	61 055	8
	.08	61 066	61 077	61 087	61 098	61 109	61 119	61 130	61 140	61 151	61 162	9
	.09	61 172	61 183	61 194	61 204	61 215	61 225	61 236	61 247	61 257	61 268	10
	4.10	.61 278	.61 289	.61 300	.61 310	.61 321	.61 331	.61 342	.61 352	.61 363	.61 374	**11**
	.11	61 384	61 395	61 405	61 416	61 426	61 437	61 448	61 458	61 469	61 479	1
	.12	61 490	61 500	61 511	61 521	61 532	61 542	61 553	61 563	61 574	61 584	2
	.13	61 595	61 606	61 616	61 627	61 637	61 648	61 658	61 669	61 679	61 690	3
	.14	61 700	61 711	61 721	61 731	61 742	61 752	61 763	61 773	61 784	61 794	4
	4.15	.61 805	.61 815	.61 826	.61 836	.61 847	.61 857	.61 868	.61 878	.61 888	.61 899	6
	.16	61 909	61 920	61 930	61 941	61 951	61 962	61 972	61 982	61 993	**62 003**	7
	.17	62 014	62 024	62 034	62 045	62 055	62 066	62 076	62 086	62 097	62 107	8
	.18	62 118	62 128	62 138	62 149	62 159	62 170	62 180	62 190	62 201	62 211	9
	.19	62 221	62 232	62 242	62 252	62 263	62 273	62 284	62 294	62 304	62 315	10
	4.20	.62 325	.62.335	.62 346	.62 356	.62 366	.62 377	.62 387	.62 397	.62 408	.62 418	**10**
	.21	62 428	62 439	62 449	62 459	62 469	62 480	62 490	62 500	62 511	62 521	1
	.22	62 531	62 542	62 552	62 562	62 572	62 583	62 593	62 603	62 613	62 624	2
	.23	62 634	62 644	62 655	62 665	62 675	62 685	62 696	62 706	62 716	62 726	3
	.24	62 737	62 747	62 757	62 767	62 778	62 788	62 798	62 808	62 818	62 829	4
	4.25	.62 839	.62 849	.62 859	.62 870	.62 880	.62 890	.62 900	.62 910	.62 921	.62 931	5
	.26	62 941	62 951	62 961	62 972	62 982	62 992	**63 002**	63 012	63 022	63 033	6
	.27	63 043	63 053	63 063	63 073	63 083	63 094	63 104	63 114	63 124	63 134	7
	.28	63 144	63 155	63 165	63 175	63 185	63 195	63 205	63 215	63 225	63 236	.8
	.29	63 246	63 256	63 266	63 276	63 286	63 296	63 306	63 317	63 327	63 337	9
	4.30	.63 347	.63 357	.63 367	.63 377	.63 387	.63 397	.63 407	.63 417	.63 428	.63 438	**10**
	.31	63 448	63 458	63 468	63 478	63 488	63 498	63 508	63 518	63 528	63 538	1
	.32	63 548	63 558	63 568	63 579	63 589	63 599	63 609	63 619	63 629	63 639	2
	.33	63 649	63 659	63 669	63 679	63 689	63 699	63 709	63 719	63 729	63 739	3
	.34	63 749	63 759	63 769	63 779	63 789	63 799	63 809	63 819	63 829	63 839	4
	4.35	.63 849	.63 859	.63 869	.63 879	.63 889	.63 899	.63 909	.63 919	.63 929	.63 939	5
	.36	63 949	63 959	63 969	63 979	63 988	63 998	**64 008**	64 018	64 028	64 038	6
	.37	64 048	64 058	64 068	64 078	64 088	64 098	64 108	64 118	64 128	64 137	7
	.38	64 147	64 157	64 167	64 177	64 187	64 197	64 207	64 217	64 227	64 237	8
	.39	64 246	64 256	64 266	64 276	64 286	64 296	64 306	64 316	64 326	64 335	9
	4.40	.64 345	.64 355	.64 365	.64 375	.64 385	.64 395	.64 404	.64 414	.64 424	.64 434	**9**
	.41	64 444	64 454	64 464	64 473	64 483	64 493	64 503	64 513	64 523	64 532	1
	.42	64 542	64 552	64 562	64 572	64 582	64 591	64 601	64 611	64 621	64 631	2
	.43	64 640	64 650	64 660	64 670	64 680	64 689	64 699	64 709	64 719	64 729	3
	.44	64 738	64 748	64 758	64 768	64 777	64 787	64 797	64 807	64 816	64 826	4
	4.45	.64 836	.64 846	.64 856	.64 865	.64 875	.64 885	.64 895	.64 904	.64 914	.64 924	5
.60	.46	64 933	64 943	64 953	64 963	64 972	64 982	64 992	**65 002**	65 011	65 021	5
.69	.47	65 031	65 040	65 050	65 060	65 070	65 079	65 089	65 099	65 108	65 118	6
	.48	65 128	65 137	65 147	65 157	65 167	65 176	65 186	65 196	65 205	65 215	7
	.49	65 225	65 234	65 244	65 254	65 263	65 273	65 283	65 292	65 302	65 312	8

No.	0	1	2	3	4	5	6	7	8	9	INTP. TAB.	
4.50	.65 321	.65 331	.65 341	.65 350	.65 360	.65 369	.65 379	.65 389	.65 398	.65 408	10	.60
.51	65 418	65 427	65 437	65 447	65 456	65 466	65 475	65 485	65 495	65 504	1	.69
.52	65 514	65 523	65 533	65 543	65 552	65 562	65 571	65 581	65 591	65 600	2	
.53	65 610	65 619	65 629	65 639	65 648	65 658	65 667	65 677	65 686	65 696	3	
.54	65 706	65 715	65 725	65 734	65 744	65 753	65 763	65 772	65 782	65 792	4	
4.55	.65 801	.65 811	.65 820	.65 830	.65 839	.65 849	.65 858	.65 868	.65 877	.65 887	5	
.56	65 896	65 906	65 916	65 925	65 935	65 944	65 954	65 963	65 973	65 982	6	
.57	65 992	**66 001**	66 011	66 020	66 030	66 039	66 049	66 058	66 068	66 077	7	
.58	66 087	66 096	66 106	66 115	66 124	66 134	66 143	66 153	66 162	66 172	8	
.59	66 181	66 191	66 200	66 210	66 219	66 229	66 238	66 247	66 257	66 266	9	
4.60	.66 276	.66 285	.66 295	.66 304	.66 314	.66 323	.66 332	.66 342	.66 351	.66 361	9	
.61	66 370	66 380	66 389	66 398	66 408	66 417	66 427	66 436	66 445	66 455	1	
.62	66 464	66 474	66 483	66 492	66 502	66 511	66 521	66 530	66 539	66 549	2	
.63	66 558	66 567	66 577	66 586	66 596	66 605	66 614	66 624	66 633	66 642	3	
.64	66 652	66 661	66 671	66 680	66 689	66 699	66 708	66 717	66 727	66 736	4	
4.65	.66 745	.66 755	.66 764	.66 773	.66 783	.66 792	.66 801	.66 811	.66 820	.66 829	5	
.66	66 839	66 848	66 857	66 867	66 876	66 885	66 894	66 904	66 913	66 922	5	
.67	66 932	66 941	66 950	66 960	66 969	66 978	66 987	66 997	**67 006**	67 015	6	
.68	67 025	67 034	67 043	67 052	67 062	67 071	67 080	67 089	67 099	67 108	7	
.69	67 117	67 127	67 136	67 145	67 154	67 164	67 173	67 182	67 191	67 201	8	
4.70	.67 210	.67 219	.67 228	.67 237	.67 247	.67 256	.67 265	.67 274	.67 284	.67 293	9	
.71	67 302	67 311	67 321	67 330	67 339	67 348	67 357	67 367	67 376	67 385	1	
.72	67 394	67 403	67 413	67 422	67 431	67 440	67 449	67 459	67 468	67 477	2	
.73	67 486	67 495	67 504	67 514	67 523	67 532	67 541	67 550	67 560	67 569	3	
.74	67 578	67 587	67 596	67 605	67 614	67 624	67 633	67 642	67 651	67 660	4	
4.75	.67 669	.67 679	.67 688	.67 697	.67 706	.67 715	.67 724	.67 733	.67 742	.67 752	5	
.76	67 761	67 770	67 779	67 788	67 797	67 806	67 815	67 825	67 834	67 843	5	
.77	67 852	67 861	67 870	67 879	67 888	67 897	67 906	67 916	67 925	67 934	6	
.78	67 943	67 952	67 961	67 970	67 979	67 988	67 997	**68 006**	68 015	68 024	7	
.79	68 034	68 043	68 052	68 061	68 070	68 079	68 088	68 097	68 106	68 115	8	
4.80	.68 124	.68 133	.68 142	.68 151	.68 160	.68 169	.68 178	.68 187	.68 196	.68 205	9	
.81	68 215	68 224	68 233	68 242	68 251	68 260	68 269	68 278	68 287	68 296	1	
.82	68 305	68 314	68 323	68 332	68 341	68 350	68 359	68 368	68 377	68 386	2	
.83	68 395	68 404	68 413	68 422	68 431	68 440	68 449	68 458	68 467	68 476	3	
.84	68 485	68 494	68 502	68 511	68 520	68 529	68 538	68 547	68 556	68 565	4	
4.85	.68 574	.68 583	.68 592	.68 601	.68 610	.68 619	.68 628	.68 637	.68 646	.68 655	5	
.86	68 664	68 673	68 681	68 690	68 699	68 708	68 717	68 726	68 735	68 744	5	
.87	68 753	68 762	68 771	68 780	68 789	68 797	68 806	68 815	68 824	68 833	6	
.88	68 842	68 851	68 860	68 869	68 878	68 886	68 895	68 904	68 913	68 922	7	
.89	68 931	68 940	68 949	68 958	68 966	68 975	68 984	68 993	**69 002**	69 011	8	
4.90	.69 020	.69 028	.69 037	.69 046	.69 055	.69 064	.69 073	.69 082	.69 090	.69 099	8	
.91	69 108	69 117	69 126	69 135	69 144	69 152	69 161	69 170	69 179	69 188	1	
.92	69 197	69 205	69 214	69 223	69 232	69 241	69 249	69 258	69 267	69 276	2	
.93	69 285	69 294	69 302	69 311	69 320	69 329	69 338	69 346	69 355	69 364	2	
.94	69 373	69 381	69 390	69 399	69 408	69 417	69 425	69 434	69 443	69 452	3	
4.95	.69 461	.69 469	.69 478	.69 487	.69 496	.69 504	.69 513	.69 522	.69 531	.69 539	4	.60
.96	69 548	69 557	69 566	69 574	69 583	69 592	69 601	69 609	69 618	69 627	5	.69
.97	69 636	69 644	69 653	69 662	69 671	69 679	69 688	69 697	69 705	69 714	6	
.98	69 723	69 732	69 740	69 749	69 758	69 767	69 775	69 784	69 793	69 801	6	
.99	69 810	69 819	69 827	69 836	69 845	69 854	69 862	69 871	69 880	69 888	7	

FIVE PLACE LOGARITHMS.

No.	0		2	3	4	5	6	7	8	9	INTP. TAB.
.70 **5.00**	.69 897	.69 906	.69 914	.69 923	.69 932	.69 940	.69 949	.69 958	.69 966	.69 975	9
.77 .01	69 984	69 992	**70 001**	70 010	70 018	70 027	70 036	70 044	70 053	70 062	1
.02	70 070	70 079	70 088	70 096	70 105	70 114	70 122	70 131	70 140	70 148	2
.03	70 157	70 165	70 174	70 183	70 191	70 200	70 209	70 217	70 226	70 234	3
.04	70 243	70 252	70 260	70 269	70 278	70 286	70 295	70 303	70 312	70 321	4
5.05	.70 329	.70 338	.70 346	.70 355	.70 364	.70 372	.70 381	.70 389	.70 398	.70 406	5
.06	70 415	70 424	70 432	70 441	70 449	70 458	70 467	70 475	70 484	70 492	5
.07	70 501	70 509	70 518	70 526	70 535	70 544	70 552	70 561	70 569	70 578	6
.08	70 586	70 595	70 603	70 612	70 621	70 629	70 638	70 646	70 655	70 663	7
.09	70 672	70 680	70 689	70 697	70 706	70 714	70 723	70 731	70 740	70 749	8
5.10	.70 757	.70 766	.70 774	.70 783	.70 791	.70 800	.70 808	.70 817	.70 825	.70 834	8
.11	70 842	70 851	70 859	70 868	70 876	70 885	70 893	70 902	70 910	70 919	1
.12	70 927	70 935	70 944	70 952	70 961	70 969	70 978	70 986	70 995	**71 003**	2
.13	71 012	71 020	71 029	71 037	71 046	71 054	71 063	71 071	71 079	71 088	2
.14	71 096	71 105	71 113	71 122	71 130	71 139	71 147	71 155	71 164	71 172	3
5.15	.71 181	.71 189	.71 198	.71 206	.71 214	.71 223	.71 231	.71 240	.71 248	.71 257	4
.16	71 265	71 273	71 282	71 290	71 299	71 307	71 315	71 324	71 332	71 341	5
.17	71 349	71 357	71 366	71 374	71 383	71 391	71 399	71 408	71 416	71 425	6
.18	71 433	71 441	71 450	71 458	71 466	71 475	71 483	71 492	71 500	71 508	6
.19	71 517	71 525	71 533	71 542	71 550	71 559	71 567	71 575	71 584	71 592	7
5.20	.71 600	.71 609	.71 617	.71 625	.71 634	.71 642	.71 650	.71 659	.71 667	.71 675	8
.21	71 684	71 692	71 700	71 709	71 717	71 725	71 734	71 742	71 750	71 759	1
.22	71 767	71 775	71 784	71 792	71 800	71 809	71 817	71 825	71 834	71 842	2
.23	71 850	71 858	71 867	71 875	71 883	71 892	71 900	71 908	71 917	71 925	2
.24	71 933	71 941	71 950	71 958	71 966	71 975	71 983	71 991	71 999	**72 008**	3
5.25	.72 016	.72 024	.72 032	.72 041	.72 049	.72 057	.72 066	.72 074	.72 082	.72 090	4
.26	72 099	72 107	72 115	72 123	72 132	72 140	72 148	72 156	72 165	72 173	5
.27	72 181	72 189	72 198	72 206	72 214	72 222	72 230	72 239	72 247	72 255	6
.28	72 263	72 272	72 280	72 288	72 296	72 304	72 313	72 321	72 329	72 337	6
.29	72 346	72 354	72 362	72 370	72 378	72 387	72 395	72 403	72 411	72 419	7
5.30	.72 428	.72 436	.72 444	.72 452	.72 460	.72 469	.72 477	.72 485	.72 493	.72 501	8
.31	72 509	72 518	72 526	72 534	72 542	72 550	72 558	72 567	72 575	72 583	1
.32	72 591	72 599	72 607	72 616	72 624	72 632	72 640	72 648	72 656	72 665	2
.33	72 673	72 681	72 689	72 697	72 705	72 713	72 722	72 730	72 738	72 746	2
.34	72 754	72 762	72 770	72 779	72 787	72 795	72 803	72 811	72 819	72 827	3
5.35	.72 835	.72 843	.72 852	.72 860	.72 868	.72 876	.72 884	.72 892	.72 900	.72 908	4
.36	72 916	72 925	72 933	72 941	72 949	72 957	72 965	72 973	72 981	72 989	5
.37	72 997	**73 006**	73 014	73 022	73 030	73 038	73 046	73 054	73 062	73 070	6
.38	73 078	73 086	73 094	73 102	73 111	73 119	73 127	73 135	73 143	73 151	6
.39	73 159	73 167	73 175	73 183	73 191	73 199	73 207	73 215	73 223	73 231	7
5.40	.73 239	.73 247	.73 255	.73 263	.73 272	.73 280	.73 288	.73 296	.73 304	.73 312	7
.41	73 320	73 328	73 336	73 344	73 352	73 360	73 368	73 376	73 384	73 392	1
.42	73 400	73 408	73 416	73 424	73 432	73 440	73 448	73 456	73 464	73 472	1
.43	73 480	73 488	73 496	73 504	73 512	73 520	73 528	73 536	73 544	73 552	2
.44	73 560	73 568	73 576	73 584	73 592	73 600	73 608	73 616	73 624	73 632	3
5.45	.73 640	.73 648	.73 656	.73 664	.73 672	.73 679	.73 687	.73 695	.73 703	.73 711	4
.70 .46	73 719	73 727	73 735	73 743	73 751	73 759	73 767	73 775	73 783	73 791	4
.77 .47	73 799	73 807	73 815	73 823	73 830	73 838	73 846	73 854	73 862	73 870	5
.48	73 878	73 886	73 894	73 902	73 910	73 918	73 926	73 933	73 941	73 949	6
.49	73 957	73 965	73 973	73 981	73 989	73 997	**74 005**	74 013	74 020	74 028	6

No.	0	1	2	3	4	5	6	7	8	9	INTP. TAB.	
5.50	.74 036	.74 044	.74 052	.74 060	.74 068	.74 076	.74 084	.74 092	.74 099	.74 107	8	.70
.51	74 115	74 123	74 131	74 139	74 147	74 155	74 162	74 170	74 178	74 186	1	.77
.52	74 194	74 202	74 210	74 218	74 225	74 233	74 241	74 249	74 257	74 265	2	
.53	74 273	74 280	74 288	74 296	74 304	74 312	74 320	74 327	74 335	74 343	2	
.54	74 351	74 359	74 367	74 374	74 382	74 390	74 398	74 406	74 414	74 421	3	
5.55	.74 429	.74 437	.74 445	.74 453	.74 461	.74 468	.74 476	.74 484	.74 492	.74 500	4	
.56	74 507	74 515	74 523	74 531	74 539	74 547	74 554	74 562	74 570	74 578	5	
.57	74 586	74 593	74 601	74 609	74 617	74 624	74 632	74 640	74 648	74 656	6	
.58	74 663	74 671	74 679	74 687	74 695	74 702	74 710	74 718	74 726	74 733	6	
.59	74 741	74 749	74 757	74 764	74 772	74 780	74 788	74 796	74 803	74 811	7	
5.60	.74 819	.74 827	.74 834	.74 842	.74 850	.74 858	.74 865	.74 873	.74 881	.74 889	7	
.61	74 896	74 904	74 912	74 920	74 927	74 935	74 943	74 950	74 958	74 966	1	
.62	74 974	74 981	74 989	74 997	**75 005**	75 012	75 020	75 028	75 035	75 043	1	
.63	75 051	75 059	75 066	75 074	75 082	75 089	75 097	75 105	75 113	75 120	2	
.64	75 128	75 136	75 143	75 151	75 159	75 166	75 174	75 182	75 189	75 197	3	
5.65	.75 205	.75 213	.75 220	.75 228	.75 236	.75 243	.75 251	.75 259	.75 266	.75 274	4	
.66	75 282	75 289	75 297	75 305	75 312	75 320	75 328	75 335	75 343	75 351	4	
.67	75 358	75 366	75 374	75 381	75 389	75 397	75 404	75 412	75 420	75 427	5	
.68	75 435	75 442	75 450	75 458	75 465	75 473	75 481	75 488	75 496	75 504	6	
.69	75 511	75 519	75 526	75 534	75 542	75 549	75 557	75 565	75 572	75 580	6	
5.70	.75 587	.75 595	.75 603	.75 610	.75 618	.75 626	.75 633	.75 641	.75 648	.75 656	8	
.71	75 664	75 671	75 679	75 686	75 694	75 702	75 709	75 717	75 724	75 732	1	
.72	75 740	75 747	75 755	75 762	75 770	75 778	75 785	75 793	75 800	75 808	2	
.73	75 815	75 823	75 831	75 838	75 846	75 853	75 861	75 868	75 876	75 884	2	
.74	75 891	75 899	75 906	75 914	75 921	75 929	75 937	75 944	75 952	75 959	3	
5.75	.75 967	.75 974	.75 982	.75 989	.75 997	**.76 005**	.76 012	.76 020	.76 027	.76 035	4	
.75	76 042	76 050	76 057	76 065	76 072	76 080	76 087	76 095	76 103	76 110	5	
.77	76 118	76 125	76 133	76 140	76 148	76 155	76 163	76 170	76 178	76 185	6	
.78	76 193	76 200	76 208	76 215	76 223	76 230	76 238	76 245	76 253	76 260	6	
.79	76 268	76 275	76 283	76 290	76 298	76 305	76 313	76 320	76 328	76 335	7	
5.80	.76 343	.76 350	.76 358	.76 365	.76 373	.76 380	.76 388	.76 395	.76 403	.76 410	7	
.81	76 418	76 425	76 433	76 440	76 448	76 455	76 462	76 470	76 477	76 485	1	
.82	76 492	76 500	76 507	76 515	76 522	76 530	76 537	76 545	76 552	76 559	1	
.83	76 567	76 574	76 582	76 589	76 597	76 604	76 612	76 619	76 626	76 634	2	
.84	76 641	76 649	76 656	76 664	76 671	76 678	76 686	76 693	76 701	76 708	3	
5.85	.76 716	.76 723	.76 730	.76 738	.76 745	.76 753	.76 760	.76 768	.76 775	.76 782	4	
.86	76 790	76 797	76 805	76 812	76 819	76 827	76 834	76 842	76 849	76 856	4	
.87	76 864	76 871	76 879	76 886	76 893	76 901	76 908	76 916	76 923	76 930	5	
.88	76 938	76 945	76 953	76 960	76 967	76 975	76 982	76 989	76 997	**77 004**	6	
.89	77 012	77 019	77 026	77 034	77 041	77 048	77 056	77 063	77 070	77 078	6	
5.90	.77 085	.77 093	.77 100	.77 107	.77 115	.77 122	.77 129	.77 137	.77 144	.77 151	7	
.91	77 159	77 166	77 173	77 181	77 188	77 195	77 203	77 210	77 217	77 225	1	
.92	77 232	77 240	77 247	77 254	77 262	77 269	77 276	77 283	77 291	77 298	1	
.93	77 305	77 313	77 320	77 327	77 335	77 342	77 349	77 357	77 364	77 371	2	
.94	77 379	77 386	77 393	77 401	77 408	77 415	77 422	77 430	77 437	77 444	3	
5.95	.77 452	.77 459	.77 466	.77 474	.77 481	.77 488	.77 495	.77 503	.77 510	.77 517	4	
.96	77 525	77 532	77 539	77 546	77 554	77 561	77 568	77 576	77 583	77 590	4	.70
.97	77 597	77 605	77 612	77 619	77 627	77 634	77 641	77 648	77 656	77 663	5	.77
.98	77 670	77 677	77 685	77 692	77 699	77 706	77 714	77 721	77 728	77 735	6	
.99	77 743	77 750	77 757	77 764	77 772	77 779	77 786	77 793	77 801	77 808	6	

	No.	0	1	2	3	4	5	6	7	8	9	INTP. TAB.
.77	**6.00**	.77 815	.77 822	.77 830	.77 837	.77 844	.77 851	.77 859	.77 866	.77 873	.77 880	**8**
.84	.01	77 887	77 895	77 902	77 909	77 916	77 924	77 931	77 938	77 945	77 952	1
	.02	77 960	77 967	77 974	77 981	77 988	77 996	**78 003**	78 010	78 017	78 025	2
	.03	78 032	78 039	78 046	78 053	78 061	78 068	78 075	78 082	78 089	78 097	2
	.04	78 104	78 111	78 118	78 125	78 132	78 140	78 147	78 154	78 161	78 168	3
	6.05	.78 176	.78 183	.78 190	.78 197	.78 204	.78 211	.78 219	.78 226	.78 233	.78 240	4
	.06	78 247	78 254	78 262	78 269	78 276	78 283	78 290	78 297	78 305	78 312	5
	.07	78 319	78 326	78 333	78 340	78 347	78 355	78 362	78 369	78 376	78 383	6
	.08	78 390	78 398	78 405	78 412	78 419	78 426	78 433	78 440	78 447	78 455	6
	.09	78 462	78 469	78 476	78 483	78 490	78 497	78 504	78 512	78 519	78 526	7
	6.10	.78 533	.78 540	.78 547	.78 554	.78 561	.78 569	.78 576	.78 583	.78 590	.78 597	**7**
	.11	78 604	78 611	78 618	78 625	78 633	78 640	78 647	78 654	78 661	78 668	1
	.12	78 675	78 682	78 689	78 696	78 704	78 711	78 718	78 725	78 732	78 739	1
	.13	78 746	78 753	78 760	78 767	78 774	78 781	78 789	78 796	78 803	78 810	2
	.14	78 817	78 824	78 831	78 838	78 845	78 852	78 859	78 866	78 873	78 880	3
	6.15	.78 888	.78 895	.78 902	.78 909	.78 916	.78 923	.78 930	.78 937	.78 944	.78 951	4
	.16	78 958	78 965	78 972	78 979	78 986	78 993	**79 000**	79 007	79 014	79 021	4
	.17	79 029	79 036	79 043	79 050	79 057	79 064	79 071	79 078	79 085	79 092	5
	.18	79 099	79 106	79 113	79 120	79 127	79 134	79 141	79 148	79 155	79 162	6
	.19	79 169	79 176	79 183	79 190	79 197	79 204	79 211	79 218	79 225	79 232	6
	6.20	.79 239	.79 246	.79 253	.79 260	.79 267	.79 274	.79 281	.79 288	.79 295	.79 302	**7**
	.21	79 309	79 316	79 323	79 330	79 337	79 344	79 351	79 358	79 365	79 372	1
	.22	79 379	79 386	79 393	79 400	79 407	79 414	79 421	79 428	79 435	79 442	1
	.23	79 449	79 456	79 463	79 470	79 477	79 484	79 491	79 498	79 505	79 511	2
	.24	79 518	79 525	79 532	79 539	79 546	79 553	79 560	79 567	79 574	79 581	3
	6.25	.79 588	.79 595	.79 602	.79 609	.79 616	.79 623	.79 630	.79 637	.79 644	.79 650	4
	.26	79 657	79 664	79 671	79 678	79 685	79 692	79 699	79 706	79 713	79 720	4
	.27	79 727	79 734	79 741	79 748	79 754	79 761	79 768	79 775	79 782	79 789	5
	.28	79 796	79 803	79 810	79 817	79 824	79 831	79 837	79 844	79 851	79 858	6
	.29	79 865	79 872	79 879	79 886	79 893	79 900	79 906	79 913	79 920	79 927	6
	6.30	.79 934	.79 941	.79 948	.79 955	.79 962	.79 969	.79 975	.79 982	.79 989	.79 996	**7**
	.31	**80 003**	80 010	80 017	80 024	80 030	80 037	80 044	80 051	80 058	80 065	1
	.32	80 072	80 079	80 085	80 092	80 099	80 106	80 113	80 120	80 127	80 134	1
	.33	80 140	80 147	80 154	80 161	80 168	80 175	80 182	80 188	80 195	80 202	2
	.34	80 209	80 216	80 223	80 229	80 236	80 243	80 250	80 257	80 264	80 271	3
	6.35	.80 277	.80 284	.80 291	.80 298	.80 305	.80 312	.80 318	.80 325	.80 332	.80 339	4
	.36	80 346	80 353	80 359	80 366	80 373	80 380	80 387	80 393	80 400	80 407	4
	.37	80 414	80 421	80 428	80 434	80 441	80 448	80 455	80 462	80 468	80 475	5
	.38	80 482	80 489	80 496	80 502	80 509	80 516	80 523	80 530	80 536	80 543	6
	.39	80 550	80 557	80 564	80 570	80 577	80 584	80 591	80 598	80 604	80 611	6
	6.40	.80 618	.80 625	.80 632	.80 638	.80 645	.80 652	.80 659	.80 665	.80 672	.80 679	**6**
	.41	80 686	80 693	80 699	80 706	80 713	80 720	80 726	80 733	80 740	80 747	1
	.42	80 754	80 760	80 767	80 774	80 781	80 787	80 794	80 801	80 808	80 814	1
	.43	80 821	80 828	80 835	80 841	80 848	80 855	80 862	80 868	80 875	80 882	2
	.44	80 889	80 895	80 902	80 909	80 916	80 922	80 929	80 936	80 943	80 949	2
	6.45	.80 956	.80 963	.80 969	.80 976	.80 983	.80 990	.80 996	**.81 003**	.81 010	.81 017	3
.77	.46	81 023	81 030	81 037	81 043	81 050	81 057	81 064	81 070	81 077	81 084	4
.84	.47	81 090	81 097	81 104	81 111	81 117	81 124	81 131	81 137	81 144	81 151	4
	.48	81 158	81 164	81 171	81 178	81 184	81 191	81 198	81 204	81 211	81 218	5
	.49	81 224	81 231	81 238	81 245	81 251	81 258	81 265	81 271	81 278	81 285	5

FIVE PLACE LOGARITHMS.

No.	0	1	2	3	4	5	6	7	8	9	INTP. TAB.	
6.50	.81 291	.81 298	.81 305	.81 311	.81 318	.81 325	.81 331	.81 338	.81 345	.81 351	6	.77
.51	81 358	81 365	81 371	81 378	81 385	81 391	81 398	81 405	81 411	81 418	1	.84
.52	81 425	81 431	81 438	81 445	81 451	81 458	81 465	81 471	81 478	81 485	1	
.53	81 491	81 498	81 505	81 511	81 518	81 525	81 531	81 538	81 544	81 551	2	
.54	81 558	81 564	81 571	81 578	81 584	81 591	81 598	81 604	81 611	81 617	2	
6.55	.81 624	.81 631	.81 637	.81 644	.81 651	.81 657	.81 664	.81 671	.81 677	.81 684	3	
.56	81 690	81 697	81 704	81 710	81 717	81 723	81 730	81 737	81 743	81 750	4	
.57	81 757	81 763	81 770	81 776	81 783	81 790	81 796	81 803	81 809	81 816	4	
.58	81 823	81 829	81 836	81 842	81 849	81 856	81 862	81 869	81 875	81 882	5	
.59	81 889	81 895	81 902	81 908	81 915	81 921	81 928	81 935	81 941	81 948	5	
6.60	.81 954	.81 961	.81 968	.81 974	.81 981	.81 987	.81 994	**.82 000**	.82 007	.82 014	7	
.61	82 020	82 027	82 033	82 040	82 046	82 053	82 060	82 066	82 073	82 079	1	
.62	82 086	82 092	82 099	82 105	82 112	82 119	82 125	82 132	82 138	82 145	1	
.63	82 151	82 158	82 164	82 171	82 178	82 184	82 191	82 197	82 204	82 210	2	
.64	82 217	82 223	82 230	82 236	82 243	82 249	82 256	82 263	82 269	82 276	3	
6.65	.82 282	.82 289	.82 295	.82 302	.82 308	.82 315	.82 321	.82 328	.82 334	.82 341	4	
.66	82 347	82 354	82 360	82 367	82 373	82 380	82 387	82 393	82 400	82 406	4	
.67	82 413	82 419	82 426	82 432	82 439	82 445	82 452	82 458	82 465	82 471	5	
.68	82 478	82 484	82 491	82 497	82 504	82 510	82 517	82 523	82 530	82 536	6	
.69	82 543	82 549	82 556	82 562	82 569	82 575	82 582	82 588	82 595	82 601	6	
6.70	.82 607	.82 614	.82 620	.82 627	.82 633	.82 640	.82 646	.82 653	.82 659	.82 666	6	
.71	82 672	82 679	82 685	82 692	82 698	82 705	82 711	82 718	82 724	82 730	1	
.72	82 737	82 743	82 750	82 756	82 763	82 769	82 776	82 782	82 789	82 795	1	
.73	82 802	82 808	82 814	82 821	82 827	82 834	82 840	82 847	82 853	82 860	2	
.74	82 866	82 872	82 879	82 885	82 892	82 898	82 905	82 911	82 918	82 924	2	
6.75	.82 930	.82 937	.82 943	.82 950	.82 956	.82 963	.82 969	.82 975	.82 982	.82 988	3	
.76	82 995	**83 001**	83 008	83 014	83 020	83 027	83 033	83 040	83 046	83 052	4	
.77	83 059	83 065	83 072	83 078	83 085	83 091	83 097	83 104	83 110	83 117	4	
.78	83 123	83 129	83 136	83 142	83 149	83 155	83 161	83 168	83 174	83 181	5	
.79	83 187	83 193	83 200	83 206	83 213	83 219	83 225	83 232	83 238	83 245	5	
6.80	.83 251	.83 257	.83 264	.83 270	.83 276	.83 283	.83 289	.83 296	.83 302	.83 308	7	
.81	83 315	83 321	83 327	83 334	83 340	83 347	83 353	83 359	83 366	83 372	1	
.82	83 378	83 385	83 391	83 398	83 404	83 410	83 417	83 423	83 429	83 436	1	
.83	83 442	83 448	83 455	83 461	83 467	83 474	83 480	83 487	83 493	83 499	2	
.84	83 506	83 512	83 518	83 525	83 531	83 537	83 544	83 550	83 556	83 563	3	
6.85	.83 569	.83 575	.83 582	.83 588	.83 594	.83 601	.83 607	.83 613	.83 620	.83 626	4	
.86	83 632	83 639	83 645	83 651	83 658	83 664	83 670	83 677	83 683	83 689	4	
.87	83 696	83 702	83 708	83 715	83 721	83 727	83 734	83 740	83 746	83 753	5	
.88	83 759	83 765	83 771	83 778	83 784	83 790	83 797	83 803	83 809	83 816	6	
.89	83 822	83 828	83 835	83 841	83 847	83 853	83 860	83 866	83 872	83 879	6	
6.90	.83 885	.83 891	.83 897	.83 904	.83 910	.83 916	.83 923	.83 929	.83 935	.83 942	6	
.91	83 948	83 954	83 960	83 967	83 973	83 979	83 985	83 992	83 998	**84 004**	1	
.92	84 011	84 017	84 023	84 029	84 036	84 042	84 048	84 055	84 061	84 067	1	
.93	84 073	84 080	84 086	84 092	84 098	84 105	84 111	84 117	84 123	84 130	2	
.94	84 136	84 142	84 148	84 155	84 161	84 167	84 173	84 180	84 186	84 192	2	
6.95	.84 198	.84 205	.84 211	.84 217	.84 223	.84 230	.84 236	.84 242	.84 248	.84 255	3	
.96	84 261	84 267	84 273	84 280	84 286	84 292	84 298	84 305	84 311	84 317	4	.77
.97	84 323	84 330	84 336	84 342	84 348	84 354	84 361	84 367	84 373	84 379	4	.84
.98	84 386	84 392	84 398	84 404	84 410	84 417	84 423	84 429	84 435	84 442	5	
.99	84 448	84 454	84 460	84 466	84 473	84 479	84 485	84 491	84 497	84 504	5	

FIVE PLACE LOGARITHMS.

	No.	0	1	2	3	4	5	6	7	8	9	INTP. TAB.
.84	**7.00**	.84 510	.84 516	.84 522	.84 528	.84 535	.84 541	.84 547	.84 553	.84 559	.84 566	7
.90	.01	84 572	84 578	84 584	84 590	84 597	84 603	84 609	84 615	84 621	84 628	1
	.02	84 634	84 640	84 646	84 652	84 658	84 665	84 671	84 677	84 683	84 689	1
	.03	84 696	84 702	84 708	84 714	84 720	84 726	84 733	84 739	84 745	84 751	2
	.04	84 757	84 763	84 770	84 776	84 782	84 788	84 794	84 800	84 807	84 813	3
	7.05	.84 819	.84 825	.84 831	.84 837	.84 844	.84 850	.84 856	.84 862	.84 868	.84 874	4
	.06	84 880	84 887	84 893	84 899	84 905	84 911	84 917	84 924	84 930	84 936	4
	.07	84 942	84 948	84 954	84 960	84 967	84 973	84 979	84 985	84 991	84 997	5
	.08	**85 003**	85 009	85 016	85 022	85 028	85 034	85 040	85 046	85 052	85 058	6
	.09	85 065	85 071	85 077	85 083	85 089	85 095	85 101	85 107	85 114	85 120	6
	7.10	.85 126	.85 132	.85 138	.85 144	.85 150	.85 156	.85 163	.85 169	.85 175	.85 181	6
	.11	85 187	85 193	85 199	85 205	85 211	85 217	85 224	85 230	85 236	85 242	1
	.12	85 248	85 254	85 260	85 266	85 272	85 278	85 285	85 291	85 297	85 303	1
	.13	85 309	85 315	85 321	85 327	85 333	85 339	85 345	85 352	85 358	85 364	2
	.14	85 370	85 376	85 382	85 388	85 394	85 400	85 406	85 412	85 418	85 425	2
	7.15	.85 431	.85 437	.85 443	.85 449	.85 455	.85 461	.85 467	.85 473	.85 479	.85 485	3
	.16	85 491	85 497	85 503	85 509	85 516	85 522	85 528	85 534	85 540	85 546	4
	.17	85 552	85 558	85 564	85 570	85 576	85 582	85 588	85 594	85 600	85 606	4
	.18	85 612	85 618	85 625	85 631	85 637	85 643	85 649	85 655	85 661	85 667	5
	.19	85 673	85 679	85 685	85 691	85 697	85 703	85 709	85 715	85 721	85 727	5
	7.20	.85 733	.85 739	.85 745	.85 751	.85 757	.85 763	.85 769	.85 775	.85 781	.85 788	6
	.21	85 794	85 800	85 806	85 812	85 818	85 824	85 830	85 836	85 842	85 848	1
	.22	85 854	85 860	85 866	85 872	85 878	85 884	85 890	85 896	85 902	85 908	1
	.23	85 914	85 920	85 926	85 932	85 938	85 944	85 950	85 956	85 962	85 968	2
	.24	·85 974	85 980	85 986	85 992	85 998	**86 004**	86 010	86 016	86 022	86 028	2
	7.25	.86 034	.86 040	.86 046	.86 052	.86 058	.86 064	.86 070	.86 076	.86 082	.86 088	3
	.26	86 094	86 100	86 106	86 112	86 118	86 124	86 130	86 136	86 141	86 147	4
	.27	86 153	86 159	86 165	86 171	86 177	86 183	86 189	86 195	86 201	86 207	4
	.28	86 213	86 219	86 225	86 231	86 237	86 243	86 249	86 255	86 261	86 267	5
	.29	86 273	86 279	86 285	86 291	86 297	86 303	86 308	86 314	86 320	86 326	5
	7.30	.86 332	.86 338	.86 344	.86 350	.86 356	.86 362	.86 368	.86 374	.86 380	.86 386	6
	.31	86 392	86 398	86 404	86 410	86 415	86 421	86 427	86 433	86 439	86 445	1
	.32	86 451	86 457	86 463	86 469	86 475	86 481	86 487	86 493	86 499	86 504	1
	.33	86 510	86 516	86 522	86 528	86 534	86 540	86 546	86 552	86 558	86 564	2
	.34	86 570	86 576	86 581	86 587	86 593	86 599	86 605	86 611	86 617	86 623	2
	7.35	.86 629	.86 635	.86 641	.86 646	.86 652	.86 658	.86 664	.86 670	.86 676	.86 682	3
	.36	86 688	86 694	86 700	86 705	86 711	86 717	86 723	86 729	86 735	86 741	4
	.37	86 747	86 753	86 759	86 764	86 770	86 776	86 782	86 788	86 794	86 800	4
	.38	86 806	86 812	86 817	86 823	86 829	86 835	86 841	86 847	86 853	86 859	5
	.39	86 864	86 870	86 876	86 882	86 888	86 894	86 900	86 906	86 911	86 917	5
	7.40	.86 923	.86 929	.86 935	.86 941	.86 947	.86 953	.86 958	.86 964	.86 970	.86 976	5
	.41	86 982	86 988	86 994	86 999	**87 005**	87 011	87 017	87 023	87 029	87 035	1
	.42	87 040	87 046	87 052	87 058	87 064	87 070	87 075	87 081	87 087	87 093	1
	.43	87 099	87 105	87 111	87 116	87 122	87 128	87 134	87 140	87 146	87 151	2
	.44	87 157	87 163	87 169	87 175	87 181	87 186	87 192	87 198	87 204	87 210	2
	7.45	.87 216	.87 221	.87 227	.87 233	.87 239	.87 245	.87 251	.87 256	.87 262	.87 268	3
.84	.46	87 274	87 280	87 286	87 291	.87 297	87 303	87 309	87 315	87 320	87 326	3
.90	.47	87 332	87 338	87 344	87 349	87 355	87 361	87 367	87 373	87 379	87 384	4
	.48	87 390	87 396	87 402	87 408	87 413	87 419	87 425	87 431	87 437	87 442	4
	.49	87 448	87 454	87 460	87 466	87 471	87 477	87 483	87 489	87 495	87 500	5

No.	0	1	2	3	4	5	6	7	8	9	INTP. TAB.	
7.50	.87 506	.87 512	.87 518	.87 523	.87 529	.87 535	.87 541	.87 547	.87 552	.87 558	**6**	**.84**
.51	87 564	87 570	87 576	87 581	87 587	87 593	87 599	87 604	87 610	87 616	1	**.90**
.52	87 622	87 628	87 633	87 639	87 645	87 651	87 656	87 662	87 668	87 674	1	
.53	87 679	87 685	87 691	87 697	87 703	87 708	87 714	87 720	87 726	87 731	2	
.54	87 737	87 743	87 749	87 754	87 760	87 766	87 772	87 777	87 783	87 789	2	
7.55	.87 795	.87 800	.87 806	.87 812	.87 818	.87 823	.87 829	.87 835	.87 841	.87 846	3	
.56	87 852	87 858	87 864	87 869	87 875	87 881	87 887	87 892	87 898	87 904	4	
.57	87 910	87 915	87 921	87 927	87 933	87 938	87 944	87 950	87 955	87 961	4	
.58	87 967	87 973	87 978	87 984	87 990	87 996	**88 001**	88 007	88 013	88 018	5	
.59	88 024	88 030	88 036	88 041	88 047	88 053	88 058	88 064	88 070	88 076	5	
7.60	.88 081	.88 087	.88 093	.88 098	.88 104	.88 110	.88 116	.88 121	.88 127	.88 133	**5**	
.61	88 138	88 144	88 150	88 156	88 161	88 167	88 173	88 178	88 184	88 190	1	
.62	88 195	88 201	88 207	88 213	88 218	88 224	88 230	88 235	88 241	88 247	1	
.63	88 252	88 258	88 264	88 270	88 275	88 281	88 287	88 292	88 298	88 304	2	
.64	88 309	88 315	88 321	88 326	88 332	88 338	88 343	88 349	88 355	88 360	2	
7.65	.88 366	.88 372	.88 377	.88 383	.88 389	.88 395	.88 400	.88 406	.88 412	.88 417	3	
.66	88 423	88 429	88 434	88 440	88 446	88 451	88 457	88 463	88 468	88 474	3	
.67	88 480	88 485	88 491	88 497	88 502	88 508	88 513	88 519	88 525	88 530	4	
.68	88 536	88 542	88 547	88 553	88 559	88 564	88 570	88 576	88 581	88 587	4	
.69	88 593	88 598	88 604	88 610	88 615	88 621	88 627	88 632	88 638	88 643	5	
7.70	.88 649	.88 655	.88 660	.88 666	.88 672	.88 677	.88 683	.88 689	.88 694	.88 700	**6**	
.71	88 705	88 711	88 717	88 722	88 728	88 734	88 739	88 745	88 750	88 756	1	
.72	88 762	88 767	88 773	88 779	88 784	88 790	88 795	88 801	88 807	88 812	1	
.73	88 818	88 824	88 829	88 835	88 840	88 846	88 852	88 857	88 863	88 868	2	
.74	88 874	88 880	88 885	88 891	88 897	88 902	88 908	88 913	88 919	88 925	2	
7.75	.88 930	.88 936	.88 941	.88 947	.88 953	.88 958	.88 964	.88 969	.88 975	.88 981	3	
.76	88 986	88 992	88 997	**89 003**	89 009	89 014	89 020	89 025	89 031	89 037	4	
.77	89 042	89 048	89 053	89 059	89 064	89 070	89 076	89 081	89 087	89 092	4	
.78	89 098	89 104	89 109	89 115	89 120	89 126	89 131	89 137	89 143	89 148	5	
.79	89 154	89 159	89 165	89 170	89 176	89 182	89 187	89 193	89 198	89 204	5	
7.80	.89 209	.89 215	.89 221	.89 226	.89 232	.89 237	.89 243	.89 248	.89 254	.89 260	**5**	
.81	89 265	89 271	89 276	89 282	89 287	89 293	89 298	89 304	89 310	89 315	1	
.82	89 321	89 326	89 332	89 337	89 343	89 348	89 354	89 360	89 365	89 371	1	
.83	89 376	89 382	89 387	89 393	89 398	89 404	89 409	89 415	89 421	89 426	2	
.84	89 432	89 437	89 443	89 448	89 454	89 459	89 465	89 470	89 476	89 481	2	
7.85	.89 487	.89 492	.89 498	.89 504	.89 509	.89 515	.89 520	.89 526	.89 531	.89 537	3	
.86	89 542	89 548	89 553	89 559	89 564	89 570	89 575	89 581	89 586	89 592	3	
.87	89 597	89 603	89 609	89 614	89 620	89 625	89 631	89 636	89 642	89 647	4	
.88	89 653	89 658	89 664	89 669	89 675	89 680	89 686	89 691	89 697	89 702	4	
.89	89 708	89 713	89 719	89 724	89 730	89 735	89 741	89 746	89 752	89 757	5	
7.90	.89 763	.89 768	.89 774	.89 779	.89 785	.89 790	.89 796	.89 801	.89 807	.89 812	**6**	
.91	89 818	89 823	89 829	89 834	89 840	89 845	89 851	89 856	89 862	89 867	1	
.92	89 873	89 878	89 883	89 889	89 894	89 900	89 905	89 911	89 916	89 922	1	
.93	89 927	89 933	89 938	89 944	89 949	89 955	89 960	89 966	89 971	89 977	2	
.94	89 982	89 988	89 993	89 998	**90 004**	90 009	90 015	90 020	90 026	90 031	2	
7.95	.90 037	.90 042	.90 048	.90 053	.90 059	.90 064	.90 069	.90 075	.90 080	.90 086	3	
.96	90 091	90 097	90 102	90 108	90 113	90 119	90 124	90 129	90 135	90 140	4	**.84**
.97	90 146	90 151	90 157	90 162	90 168	90 173	90 179	90 184	90 189	90 195	4	**.90**
.98	90 200	90 206	90 211	90 217	90 222	90 227	90 233	90 238	90 244	90 249	5	
.99	90 255	90 260	90 266	90 271	90 276	90 282	90 287	90 293	90 298	90 304	5	

	No.	0	1	2	3	4	5	6	7	8	9	INTP. TAB.
.90	**8.00**	.90 309	.90 314	.90 320	.90 325	.90 331	.90 336	.90 342	.90 347	.90 352	.90 358	5
.95	.01	90 363	90 369	90 374	90 380	90 385	90 390	90 396	90 401	90 407	90 412	1
	.02	90 417	90 423	90 428	90 434	90 439	90 445	90 450	90 455	90 461	90 466	1
	.03	90 472	90 477	90 482	90 488	90 493	90 499	90 504	90 509	90 515	90 520	2
	.04	90 526	90 531	90 536	90 542	90 547	90 553	90 558	90 563	90 569	90 574	2
	8.05	.90 580	.90 585	.90 590	.90 596	.90 601	.90 607	.90 612	.90 617	.90 623	.90 628	3
	.06	90 634	90 639	90 644	90 650	90 655	90 660	90 666	90 671	90 677	90 682	3
	.07	90 687	90 693	90 698	90 703	90 709	90 714	90 720	90 725	90 730	90 736	4
	.08	90 741	90 747	90 752	90 757	90 763	90 768	90 773	90 779	90 784	90 789	4
	.09	90 795	90 800	90 806	90 811	90 816	90 822	90 827	90 832	90 838	90 843	5
	8.10	.90 849	.90 854	.90 859	.90 865	.90 870	.90 875	.90 881	.90 886	.90 891	.90 897	6
	.11	90 902	90 907	90 913	90 918	90 924	90 929	90 934	90 940	90 945	90 950	1
	.12	90 956	90 961	90 966	90 972	90 977	90 982	90 988	90 993	90 998	**91 004**	1
	.13	91 009	91 014	91 020	91 025	91 030	91 036	91 041	91 046	91 052	91 057	2
	.14	91 062	91 068	91 073	91 078	91 084	91 089	91 094	91 100	91 105	91 110	2
	8.15	.91 116	.91 121	.91 126	.91 132	.91 137	.91 142	.91 148	.91 153	.91 158	.91 164	3
	.16	91 169	91 174	91 180	91 185	91 190	91 196	91 201	91 206	91 212	91 217	4
	.17	91 222	91 228	91 233	91 238	91 243	91 249	91 254	91 259	91 265	91 270	4
	.18	91 275	91 281	91 286	91 291	91 297	91 302	91 307	91 312	91 318	91 323	5
	.19	91 328	91 334	91 339	91 344	91 350	91 355	91 360	91 365	91 371	91 376	5
	8.20	.91 381	.91 387	.91 392	.91 397	.91 402	.91 408	.91 413	.91 418	.91 424	.91 429	5
	.21	91 434	91 440	91 445	91 450	91 455	91 461	91 466	91 471	91 477	91 482	1
	.22	91 487	91 492	91 498	91 503	91 508	91 514	91 519	91 524	91 529	91 535	1
	.23	91 540	91 545	91 551	91 556	91 561	91 566	91 572	91 577	91 582	91 587	2
	.24	91 593	91 598	91 603	91 609	91 614	91 619	91 624	91 630	91 635	91 640	2
	8.25	.91 645	.91 651	.91 656	.91 661	.91 666	.91 672	.91 677	.91 682	.91 687	.91 693	3
	.26	91 698	91 703	91 709	91 714	91 719	91 724	91 730	91 735	91 740	91 745	3
	.27	91 751	91 756	91 761	91 766	91 772	91 777	91 782	91 787	91 793	91 798	4
	.28	91 803	91 808	91 814	91 819	91 824	91 829	91 834	91 840	91 845	91 850	4
	.29	91 855	91 861	91 866	91 871	91 876	91 882	91 887	91 892	91 897	91 903	5
	8.30	.91 908	.91 913	.91 918	.91 924	.91 929	.91 934	.91 939	.91 944	.91 950	.91 955	6
	.31	91 960	91 965	91 971	91 976	91 981	91 986	91 991	91 997	**92 002**	92 007	1
	.32	92 012	92 018	92 023	92 028	92 033	92 038	92 044	92 049	92 054	92 059	1
	.33	92 065	92 070	92 075	92 080	92 085	92 091	92 096	92 101	92 106	92 111	2
	.34	92 117	92 122	92 127	92 132	92 137	92 143	92 148	92.153	92 158	92 163	2
	8.35	.92 169	.92 174	.92 179	.92 184	.92 189	.92 195	.92 200	.92 205	.92 210	.92 215	3
	.36	92 221	92 226	92 231	92 236	92 241	92 247	92 252	92 257	92 262	92 267	4
	.37	92 273	92 278	92 283	92 288	92 293	92 298	92 304	92 309	92 314	92 319	4
	.38	92 324	92 330	92 335	92 340	92 345	92 350	92 355	92 361	92 366	92 371	5
	.39	92 376	92 381	92 387	92 392	92 397	92 402	92 407	92 412	92 418	92 423	5
	8.40	.92 428	.92 433	.92 438	.92 443	.92 449	.92 454	.92 459	.92 464	.92 469	.92 474	5
	.41	92 480	92 485	92 490	92 495	92 500	92 505	92 511	92 516	92 521	92 526	1
	.42	92 531	92 536	92 542	92 547	92 552	92 557	92 562	92 567	92 572	92 578	1
	.43	92 583	92 588	92 593	92 598	92 603	92 609	92 614	92 619	92 624	92 629	2
	.44	92 634	92 639	92 645	92 650	92 655	92 660	92 665	92 670	92 675	92 681	2
	8.45	.92 686	.92 691	.92 696	.92 701	.92 706	.92 711	.92 716	.92 722	.92 727	.92 732	3
.90	.46	92 737	92 742	92 747	92 752	92 758	92 763	92 768	92 773	92 778	92 783	3
.95	.47	92 788	92 793	92 799	92 804	92 809	92 814	92 819	92 824	92 829	92 834	4
	.48	92 840	92 845	92 850	92 855	92 860	92 865	92 870	92 875	92 881	92 886	4
	.49	92 891	92 896	92 901	92 906	92 911	92.916	92 921	92 927	92 932	92 937	5

FIVE PLACE LOGARITHMS.

No.	0	1	2	3	4	5	6	7	8	9	INTP. TAB.	
8.50	.92 942	.92 947	.92 952	.92 957	.92 962	.92 967	.92 973	.92 978	.92 983	.92 988	6	.90
.51	92 993	92 998	**93 003**	93 008	93 013	93 018	93 024	93 029	93 034	93 039	1	.95
.52	93 044	93 049	93 054	93 059	93 064	93 069	93 075	93 080	93 085	93 090	1	
.53	93 095	93 100	93 105	93 110	93 115	93 120	93 125	93 131	93 136	93 141	2	
.54	93 146	93 151	93 156	93 161	93 166	93 171	93 176	93 181	93 186	93 192	2	
8.55	.93 197	.93 202	.93 207	.93 212	.93 217	.93 222	.93 227	.93 232	.93 237	.93 242	3	
.56	93 247	93 252	93 258	93 263	93 268	93 273	93 278	93 283	93 288	93 293	4	
.57	93 298	93 303	93 308	93 313	93 318	93 323	93 328	93 334	93 339	93 344	4	
.58	93 349	93 354	93 359	93 364	93 369	93 374	93 379	93 384	93 389	93 394	5	
.59	93 399	93 404	93 409	93 414	93 420	93 425	93 430	93 435	93 440	93 445	5	
8.60	.93 450	.93 455	.93 460	.93 465	.93 470	.93 475	.93 480	.93 485	.93 490	.93 495	5	
.61	93 500	93 505	93 510	93 515	93 520	93 526	93 531	93 536	93 541	93 546	1	
.62	93 551	93 556	93 561	93 566	93 571	93 576	93 581	93 586	93 591	93 596	1	
.63	93 601	93 606	93 611	93 616	93 621	93 626	93 631	93 636	93 641	93 646	2	
.64	93 651	93 656	93 661	93 666	93 671	93 676	93 682	93 687	93 692	93 697	2	
8.65	.93 702	.93 707	.93 712	.93 717	.93 722	.93 727	.93 732	.93 737	.93 742	.93 747	3	
.66	93 752	93 757	93 762	93 767	93 772	93 777	93 782	93 787	93 792	93 797	3	
.67	93 802	93 807	93 812	93 817	93 822	93 827	93 832	93 837	93 842	93 847	4	
.68	93 852	93 857	93 862	93 867	93 872	93 877	93 882	93 887	93 892	93 897	4	
.69	93 902	93 907	93 912	93 917	93 922	93 927	93 932	93 937	93 942	93 947	5	
8.70	.93 952	.93 957	.93 962	.93 967	.93 972	.93 977	.93 982	.93 987	.93 992	.93 997	4	
.71	**94 002**	94 007	94 012	94 017	94 022	94 027	94 032	94 037	94 042	94 047	0	
.72	94 052	94 057	94 062	94 067	94 072	94 077	94 082	94 086	94 091	94 096	1	
.73	94 101	94 106	94 111	94 116	94 121	94 126	94 131	94 136	94 141	94 146	1	
.74	94 151	94 156	94 161	94 166	94 171	94 176	94 181	94 186	94 191	94 196	2	
8.75	.94 201	.94 206	.94 211	.94 216	.94 221	.94 226	.94 231	.94 236	.94 240	.94 245	2	
.76	94 250	94 255	94 260	94 265	94 270	94 275	94 280	94 285	94 290	94 295	2	
.77	94 300	94 305	94 310	94 315	94 320	94 325	94 330	94 335	94 340	94 345	3	
.78	94 349	94 354	94 359	94 364	94 369	94 374	94 379	94 384	94 389	94 394	3	
.79	94 399	94 404	94 409	94 414	94 419	94 424	94 429	94 433	94 438	94 443	4	
8.80	.94 448	.94 453	.94 458	.94 463	.94 468	.94 473	.94 478	.94 483	.94 488	.94 493	5	
.81	94 498	94 503	94 507	94 512	94 517	94 522	94 527	94 532	94 537	94 542	1	
.82	94 547	94 552	94 557	94 562	94 567	94 571	94 576	94 581	94 586	94 591	1	
.83	94 596	94 601	94 606	94 611	94 616	94 621	94 626	94 630	94 635	94 640	2	
.84	94 645	94 650	94 655	94 660	94 665	94 670	94 675	94 680	94 685	94 689	2	
8.85	.94 694	.94 699	.94 704	.94 709	.94 714	.94 719	.94 724	.94 729	.94 734	.94 738	3	
.86	94 743	94 748	94 753	94 758	94 763	94 768	94 773	94 778	94 783	94 787	3	
.87	94 792	94 797	94 802	94 807	94 812	94 817	94 822	94 827	94 832	94 836	4	
.88	94 841	94 846	94 851	94 856	94 861	94 866	94 871	94 876	94 880	94 885	4	
.89	94 890	94 895	94 900	94 905	94 910	94 915	94 919	94 924	94 929	94 934	5	
8.90	.94 939	.94 944	.94 949	.94 954	.94 959	.94 963	.94 968	.94 973	.94 978	.94 983	4	
.91	94 988	94 993	94 998	**95 002**	95 007	95 012	95 017	95 022	95 027	95 032	0	
.92	95 036	95 041	95 046	95 051	95 056	95 061	95 066	95 071	95 075	95 080	1	
.93	95 085	95 090	95 095	95 100	95 105	95 109	95 114	95 119	95 124	95 129	1	
.94	95 134	95 139	95 143	95 148	95 153	95 158	95 163	95 168	95 173	95 177	2	
8.95	.95 182	.95 187	.95 192	.95 197	.95 202	.95 207	.95 211	.95 216	.95 221	.95 226	2	
.96	95 231	95 236	95 240	95 245	95 250	95 255	95 260	95 265	95 270	95 274	2	.90
.97	95 279	95 284	95 289	95 294	95 299	95 303	95 308	95 313	95 318	95 323	3	.95
.98	95 328	95 332	95 337	95 342	95 347	95 352	95 357	95 361	95 366	95 371	3	
.99	95 376	95 381	95 386	95 390	95 395	95 400	95 405	95 410	95 415	95 419	4	

FIVE PLACE LOGARITHMS.

	No.	0	1	2	3	4	5	6	7	8	9	INTP. TAB.
.95	**9.00**	.95 424	.95 429	.95 434	.95 439	.95 444	.95 448	.95 453	.95 458	.95 463	.95 468	5
.99	.01	95 472	95 477	95 482	95 487	95 492	95 497	95 501	95 506	95 511	95 516	1
	.02	95 521	95 525	95 530	95 535	95 540	95 545	95 550	95 554	95 559	95 564	1
	.03	95 569	95 574	95 578	95 583	95 588	95 593	95 598	95 602	95 607	95 612	2
	.04	95 617	95 622	95 626	95 631	95 636	95 641	95 646	95 650	95 655	95 660	2
	9.05	.95 665	.95 670	.95 674	.95 679	.95 684	.95 689	.95 694	.95 698	.95 703	.95 708	3
	.06	95 713	95 718	95 722	95 727	95 732	95 737	95 742	95 746	95 751	95 756	3
	.07	95 761	95 766	95 770	95 775	95 780	95 785	95 789	95 794	95 799	95 804	4
	.08	95 809	95 813	95 818	95 823	95 828	95 832	95 837	95 842	95 847	95 852	4
	.09	95 856	95 861	95 866	95 871	95 875	95 880	95 885	95 890	95 895	95 899	5
	9.10	.95 904	.95 909	.95 914	.95 918	.95 923	.95 928	.95 933	.95 938	.95 942	.95 947	4
	.11	95 952	95 957	95 961	95 966	95 971	95 976	95 980	95 985	95 990	95 995	0
	.12	95 999	**96 004**	96 009	96 014	96 019	96 023	96 028	96 033	96 038	96 042	1
	.13	96 047	96 052	96 057	96 061	96 066	96 071	96 076	96 080	96 085	96 090	1
	.14	96 095	96 099	96 104	96 109	96 114	96 118	96 123	96 128	96 133	96 137	2
	9.15	.96 142	.96 147	.96 152	.96 156	.96 161	.96 166	.96 171	.96 175	.96 180	.96 185	2
	.16	96 190	96 194	96 199	96 204	96 209	96 213	96 218	96 223	96 227	96 232	2
	.17	96 237	96 242	96 246	96 251	96 256	96 261	96 265	96 270	96 275	96 280	3
	.18	96 284	96 289	96 294	96 298	96 303	96 308	96 313	96 317	96 322	96 327	3
	.19	96 332	96 336	96 341	96 346	96 350	96 355	96 360	96 365	96 369	96 374	4
	9.20	.96 379	.96 384	.96 388	.96 393	.96 398	.96 402	.96 407	.96 412	.96 417	.96 421	5
	.21	96 426	96 431	96 435	96 440	96 445	96 450	96 454	96 459	96 464	96 468	1
	.22	96 473	96 478	96 483	96 487	96 492	96 497	96 501	96 506	96 511	96 515	1
	.23	96 520	96 525	96 530	96 534	96 539	96 544	96 548	96 553	96 558	96 562	2
	.24	96 567	96 572	96 577	96 581	96 586	96 591	96 595	96 600	96 605	96 609	2
	9.25	.96 614	.96 619	.96 624	.96 628	.96 633	.96 638	.96 642	.96 647	.96 652	.96 656	3
	.26	96 661	96 666	96 670	96 675	96 680	96 685	96 689	96 694	96 699	96 703	3
	.27	96 708	96 713	96 717	96 722	96 727	96 731	96 736	96 741	96 745	96 750	4
	.28	96 755	96 759	96 764	96 769	96 774	96 778	96 783	96 788	96 792	96 797	4
	.29	96 802	96 806	96 811	96 816	96 820	96 825	96 830	96 834	96 839	96 844	5
	9.30	.96 848	.96 853	.96 858	.96 862	.96 867	.96 872	.96 876	.96 881	.96 886	.96 890	4
	.31	96 895	96 900	96 904	96 909	96 914	96 918	96 923	96 928	96 932	96 937	0
	.32	96 942	96 946	96 951	96 956	96 960	96 965	96 970	96 974	96 979	96 984	1
	.33	96 988	96 993	96 997	**97 002**	97 007	97 011	97 016	97 021	97 025	97 030	1
	.34	97 035	97 039	97 044	97 049	97 053	97 058	97 063	97 067	97 072	97 077	2
	9.35	.97 081	.97 086	.97 090	.97 095	.97 100	.97 104	.97 109	.97 114	.97 118	.97 123	2
	.36	97 128	97 132	97 137	97 142	97 146	97 151	97 155	97 160	97 165	97 169	2
	.37	97 174	97 179	97 183	97 188	97 192	97 197	97 202	97 206	97 211	97 216	3
	.38	97 220	97 225	97 230	97 234	97 239	97 243	97 248	97 253	97 257	97 262	3
	.39	97 267	97 271	97 276	97 280	97 285	97 290	97 294	97 299	97 304	97 308	4
	9.40	.97 313	.97 317	.97 322	.97 327	.97 331	.97 336	.97 340	.97 345	.97 350	.97 354	5
	.41	97 359	97 364	97 368	97 373	97 377	97 382	97 387	97 391	97 396	97 400	1
	.42	97 405	97 410	97 414	97 419	97 424	97 428	97 433	97 437	97 442	97 447	1
	.43	97 451	97 456	97 460	97 465	97 470	97 474	97 479	97 483	97 488	97 493	2
	.44	97 497	97 502	97 506	97 511	97 516	97 520	97 525	97 529	97 534	97 539	2
	9.45	.97 543	.97 548	.97 552	.97 557	.97 562	.97 566	.97 571	.97 575	.97 580	.97 585	3
.95	.46	97 589	97 594	97 598	97 603	97 607	97 612	97 617	97 621	97 626	97 630	3
.99	.47	97 635	97 640	97 644	97 649	97 653	97 658	97 663	97 667	97 672	97 676	4
	.48	97 681	97 685	97 690	97 695	97 699	97 704	97 708	97 713	97 717	97 722	4
	.49	97 727	97 731	97 736	97 740	97 745	97 749	97 754	97 759	97 763	97 768	5

No.	0	1	2	3	4	5	6	7	8	9	INTP. TAB.	
9.50	.97 772	.97 777	.97 782	.97 786	.97 791	.97 795	.97 800	.97 804	.97 809	.97 813	4	.95
.51	97 818	97 823	97 827	97 832	97 836	97 841	97 845	97 850	97 855	97 859	0	.99
.52	97 864	97 868	97 873	97 877	97 882	97 886	97 891	97 896	97 900	97 905	1	
.53	97 909	97 914	97 918	97 923	97 928	97 932	97 937	97 941	97 946	97 950	1	
.54	97 955	97 959	97 964	97 968	97 973	97 978	97 982	97 987	97 991	97 996	2	
9.55	**.98 000**	.98 005	.98 009	.98 014	.98 019	.98 023	.98 028	.98 032	.98 037	.98 041	2	
.56	98 046	98 050	98 055	98 059	98 064	98 068	98 073	98 078	98 082	98 087	2	
.57	98 091	98 096	98 100	98 105	98 109	98 114	98 118	98 123	98 127	98 132	3	
.58	98 137	98 141	98 146	98 150	98 155	98 159	98 164	98 168	98 173	98 177	3	
.59	98 182	98 186	98 191	98 195	98 200	98 204	98 209	98 214	98 218	98 223	4	
9.60	.98 227	.98 232	.98 236	.98 241	.98 245	.98 250	.98 254	.98 259	.98 263	.98 268	5	
.61	98 272	98 277	98 281	98 286	98 290	98 295	98 299	98 304	98 308	98 313	1	
.62	98 318	98 322	98 327	98 331	98 336	98 340	98 345	98 349	98 354	98 358	1	
.63	98 363	98 367	98 372	98 376	98 381	98 385	98 390	98 394	98 399	98 403	2	
.64	98 408	98 412	98 417	98 421	98 426	98 430	98 435	98 439	98 444	98 448	2	
9.65	.98 453	.98 457	.98 462	.98 466	.98 471	.98 475	.98 480	.98 484	.98 489	.98 493	3	
.66	98 498	98 502	98 507	98 511	98 516	98 520	98 525	98 529	98 534	98 538	3	
.67	98 543	98 547	98 552	98 556	98 561	98 565	98 570	98 574	98 579	98 583	4	
.68	98 588	98 592	98 597	98 601	98 605	98 610	98 614	98 619	98 623	98 628	4	
.69	98 632	98 637	98 641	98 646	98 650	98 655	98 659	98 664	98 668	98 673	5	
9.70	.98 677	.98 682	.98 686	.98 691	.98 695	.98 700	.98 704	.98 709	.98 713	.98 717	4	
.71	98 722	98 726	98 731	98 735	98 740	98 744	98 749	98 753	98 758	98 762	0	
.72	98 767	98 771	98 776	98 780	98 784	98 789	98 793	98 798	98 802	98 807	1	
.73	98 811	98 816	98 820	98 825	98 829	98 834	98 838	98 843	98 847	98 851	1	
.74	98 856	98 860	98 865	98 869	98 874	98 878	98 883	98 887	98 892	98 896	2	
9.75	.98 900	.98 905	.98 909	.98 914	.98 918	.98 923	.98 927	.98 932	.98 936	.98 941	2	
.76	98 945	98 949	98 954	98 958	98 963	98 967	98 972	98 976	98 981	98 985	2	
.77	98 989	98 994	98 998	**99 003**	99 007	99 012	99 016	99 021	99 025	99 029	3	
.78	99 034	99 038	99 043	99 047	99 052	99 056	99 061	99 065	99 069	99 074	3	
.79	99 078	99 083	99 087	99 092	99 096	99 100	99 105	99 109	99 114	99 118	4	
9.80	.99 123	.99 127	.99 131	.99 136	.99 140	.99 145	.99 149	.99 154	.99 158	.99 162	5	
.81	99 167	99 171	99 176	99 180	99 185	99 189	99 193	99 198	99 202	99 207	1	
.82	99 211	99 216	99 220	99 224	99 229	99 233	99 238	99 242	99 247	99 251	1	
.83	99 255	99 260	99 264	99 269	99 273	99 277	99 282	99 286	99 291	99 295	2	
.84	99 300	99 304	99 308	99 313	99 317	99 322	99 326	99 330	99 335	99 339	2	
9.85	.99 344	.99 348	.99 352	.99 357	.99 361	.99 366	.99 370	.99 374	.99 379	.99 383	3	
.86	99 388	99 392	99 396	99 401	99 405	99 410	99 414	99 419	99 423	99 427	3	
.87	99 432	99 436	99 441	99 445	99 449	99 454	99 458	99 463	99 467	99 471	4	
.88	99 476	99 480	99 484	99 489	99 493	99 498	99 502	99 506	99 511	99 515	4	
.89	99 520	99 524	99 528	99 533	99 537	99 542	99 546	99 550	99 555	99 559	5	
9.90	.99 564	.99 568	.99 572	.99 577	.99 581	.99 585	.99 590	.99 594	.99 599	.99 603	4	
.91	99 607	99 612	99 616	99 621	99 625	99 629	99 634	99 638	99 642	99 647	0	
.92	99 651	99 656	99 660	99 664	99 669	99 673	99 677	99 682	99 686	99 691	1	
.93	99 695	99 699	99 704	99 708	99 712	99 717	99 721	99 726	99 730	99 734	1	
.94	99 739	99 743	99 747	99 752	99 756	99 760	99 765	99 769	99 774	99 778	2	
9.95	.99 782	.99 787	.99 791	.99 795	.99 800	.99 804	.99 808	.99 813	.99 817	.99 822	2	
.96	99 826	99 830	99 835	99 839	99 843	99 848	99 852	99 856	99 861	99 865	2	.95
.97	99 876	99 874	99 878	99 883	99 887	99 891	99 896	99 900	99 904	99 909	3	.99
.98	99 913	99 917	99 922	99 926	99 930	99 935	99 939	99 944	99 948	99 952	3	
.99	99 957	99 961	99 965	99 970	99 974	99 978	99 983	99 987	99 991	99 996	4	

SQUARE ROOTS AND SQUARES.

Note. The table gives roots directly, squares by inverse interpolation.

No.	0	1	2	3	4	5	6	7	8	9	Interpola. for Thousandths.	
1.0	1.000	1.005	1.010	1.015	1.020	1.025	1.030	1.034	1.039	1.044	**5**	**4**
.1	1.049	1.054	1.058	1.063	1.068	1.072	1.077	1.082	1.086	1.091	1	0
.2	1.095	**1.100**	1.105	1.109	1.114	1.118	1.122	1.127	1.131	1.136	1	1
.3	1.140	1.145	1.149	1.153	1.158	1.162	1.166	1.170	1.175	1.179	2	1
.4	1.183	1.187	1.192	1.196	**1.200**	1.204	1.208	1.212	1.217	1.221	2	2
1.5	1.225	1.229	1.233	1.237	1.241	1.245	1.249	1.253	1.257	1.261	3	2
.6	1.265	1.269	1.273	1.277	1.281	1.285	1.288	1.292	1.296	**1.300**	3	2
.7	1.304	1.308	1.311	1.315	1.319	1.323	1.327	1.330	1.334	1.338	4	3
.8	1.342	1.345	1.349	1.353	1.356	1.360	1.364	1.367	1.371	1.375	4	3
.9	1.378	1.382	1.386	1.389	1.393	1.396	**1.400**	1.404	1.407	1.411	5	4
2.0	1.414	1.418	1.421	1.425	1.428	1.432	1.435	1.439	1.442	1.446	**4**	**3**
.1	1.449	1.453	1.456	1.459	1.463	1.466	1.470	1.473	1.476	1.480	0	0
.2	1.483	1.487	1.490	1.493	1.497	**1.500**	1.503	1.507	1.510	1.513	1	1
.3	1.517	1.520	1.523	1.526	1.530	1.533	1.536	1.539	1.543	1.546	1	1
.4	1.549	1.552	1.556	1.559	1.562	1.565	1.568	1.572	1.575	1.578	2	1
2.5	1.581	1.584	1.587	1.591	1.594	1.597	**1.600**	1.603	1.606	1.609	2	2
.6	1.612	1.616	1.619	1.622	1.625	1.628	1.631	1.634	1.637	1.640	2	2
.7	1.643	1.646	1.649	1.652	1.655	1.658	1.661	1.664	1.667	1.670	3	2
.8	1.673	1.676	1.679	1.682	1.685	1.688	1.691	1.694	1.697	**1.700**	3	2
.9	1.703	1.706	1.709	1.712	1.715	1.718	1.720	1.723	1.726	1.729	4	3
3.0	1.732	1.735	1.738	1.741	1.744	1.746	1.749	1.752	1.755	1.758	**3**	**2**
.1	1.761	1.764	1.766	1.769	1.772	1.775	1.778	1.780	1.783	1.786	0	0
.2	1.789	1.792	1.794	1.797	**1.800**	1.803	1.806	1.808	1.811	1.814	1	0
.3	1.817	1.819	1.822	1.825	1.828	1.830	1.833	1.836	1.838	1.841	1	1
.4	1.844	1.847	1.849	1.852	1.855	1.857	1.860	1.863	1.865	1.868	1	1
3.5	1.871	1.873	1.876	1.879	1.881	1.884	1.887	1.889	1.892	1.895	2	1
.6	1.897	**1.900**	1.903	1.905	1.908	1.910	1.913	1.916	1.918	1.921	2	1
.7	1.924	1.926	1.929	1.931	1.934	1.936	1.939	1.942	1.944	1.947	2	1
.8	1.949	1.952	1.954	1.957	1.960	1.962	1.965	1.967	1.970	1.972	2	2
.9	1.975	1.977	1.980	1.982	1.985	1.987	1.990	1.992	1.995	1.997	3	2
4.0	**2.000**	2.002	2.005	2.007	2.010	2.012	2.015	2.017	2.020	2.022	**3**	**2**
.1	2.025	2.027	2.030	2.032	2.035	2.037	2.040	2.042	2.045	2.047	0	0
.2	2.049	2.052	2.054	2.057	2.059	2.062	2.064	2.066	2.069	2.071	1	0
.3	2.074	2.076	2.078	2.081	2.083	2.086	2.088	2.090	2.093	2.095	1	1
.4	2.098	**2.100**	2.102	2.105	2.107	2.110	2.112	2.114	2.117	2.119	1	1
4.5	2.121	2.124	2.126	2.128	2.131	2.133	2.135	2.138	2.140	2.142	2	1
.6	2.145	2.147	2.149	2.152	2.154	2.156	2.159	2.161	2.163	2.166	2	1
.7	2.168	2.170	2.173	2.175	2.177	2.179	2.182	2.184	2.186	2.189	2	1
.8	2.191	2.193	2.195	2.198	**2.200**	2.202	2.205	2.207	2.209	2.211	2	2
.9	2.214	2.216	2.218	2.220	2.223	2.225	2.227	2.229	2.232	2.234	3	2

No.	0	1	2	3	4	5	6	7	8	9	Interpola. for Thousandths.	
5.0	2.236	2.238	2.241	2.243	2.245	2.247	2.249	2.252	2.254	2.256	**3**	**2**
.1	2.258	2.261	2.263	2.265	2.267	2.269	2.272	2.274	2.276	2.278	0	0
.2	2.280	2.283	2.285	2.287	2.289	2.291	2.293	2.296	2.298	**2.300**	1	0
.3	2.302	2.304	2.307	2.309	2.311	2.313	2.315	2.317	2.319	2.322	1	1
.4	2.324	2.326	2.328	2.330	2.332	2.335	2.337	2.339	2.341	2.343	1	1
5.5	2.345	2.347	2.349	2.352	2.354	2.356	2.358	2.360	2.362	2.364	2	1
.6	2.366	2.369	2.371	2.373	2.375	2.377	2.379	2.381	2.383	2.385	2	1
.7	2.387	2.390	2.392	2.394	2.396	2.398	**2.400**	2.402	2.404	2.406	2	1
.8	2.408	2.410	2.412	2.415	2.417	2.419	2.421	2.423	2.425	2.427	2	2
.9	2.429	2.431	2.433	2.435	2.437	2.439	2.441	2.443	2.445	2.447	3	2
6.0	2.449	2.452	2.454	2.456	2.458	2.460	2.462	2.464	2.466	2.468	**3**	**2**
.1	2.470	2.472	2.474	2.476	2.478	2.480	2.482	2.484	2.486	2.488	0	0
.2	2.490	2.492	2.494	2.496	2.498	**2.500**	2.502	2.504	2.506	2.508	1	0
.3	2.510	2.512	2.514	2.516	2.518	2.520	2.522	2.524	2.526	2.528	1	1
.4	2.530	2.532	2.534	2.536	2.538	2.540	2.542	2.544	2.546	2.548	1	1
6.5	2.550	2.551	2.553	2.555	2.557	2.559	2.561	2.563	2.565	2.567	2	1
.6	2.569	2.571	2.573	2.575	2.577	2.579	2.581	2.583	2.585	2.587	2	1
.7	2.588	2.590	2.592	2.594	2.596	2.598	**2.600**	2.602	2.604	2.606	2	1
.8	2.608	2.610	2.612	2.613	2.615	2.617	2.619	2.621	2.623	2.625	2	2
.9	2.627	2.629	2.631	2.632	2.634	2.636	2.638	2.640	2.642	2.644	3	2
7.0	2.646	2.648	2.650	2.651	2.653	2.655	2.657	2.659	2.661	2.663	**2**	**1**
.1	2.665	2.666	2.668	2.670	2.672	2.674	2.676	2.678	2.680	2.681	0	0
.2	2.683	2.685	2.687	2.689	2.691	2.693	2.694	2.696	2.698	**2.700**	0	0
.3	2.702	2.704	2.706	2.707	2.709	2.711	2.713	2.715	2.717	2.718	1	0
.4	2.720	2.722	2.724	2.726	2.728	2.729	2.731	2.733	2.735	2.737	1	0
7.5	2.739	2.740	2.742	2.744	2.746	2.748	2.750	2.751	2.753	2.755	1	1
.6	2.757	2.759	2.760	2.762	2.764	2.766	2.768	2.769	2.771	2.773	1	1
.7	2.775	2.777	2.778	2.780	2.782	2.784	2.786	2.787	2.789	2.791	1	1
.8	2.793	2.795	2.796	2.798	**2.800**	2.802	2.804	2.805	2.807	2.809	2	1
.9	2.811	2.812	2.814	2.816	2.818	2.820	2.821	2.823	2.825	2.827	2	1
8.0	2.828	2.830	2.832	2.834	2.835	2.837	2.839	2.841	2.843	2.844	**2**	**1**
.1	2.846	2.848	2.850	2.851	2.853	2.855	2.857	2.858	2.860	2.862	0	0
.2	2.864	2.865	2.867	2.869	2.871	2.872	2.874	2.876	2.877	2.879	0	0
.3	2.881	2.883	2.884	2.886	2.888	2.890	2.891	2.893	2.895	2.897	1	0
.4	2.898	**2.900**	2.902	2.903	2.905	2.907	2.909	2.910	2.912	2.914	1	0
8.5	2.915	2.917	2.919	2.921	2.922	2.924	2.926	2.927	2.929	2.931	1	1
.6	2.933	2.934	2.936	2.938	2.939	2.941	2.943	2.944	2.946	2.948	1	1
.7	2.950	2.951	2.953	2.955	2.956	2.958	2.960	2.961	2.963	2.965	1	1
.8	2.966	2.968	2.970	2.972	2.973	2.975	2.977	2.978	2.980	2.982	2	1
.9	2.983	2.985	2.987	2.988	2.990	2.992	2.993	2.995	2.997	2.998	2	1
9.0	**3.000**	3.002	3.003	3.005	3.007	3.008	3.010	3.012	3.013	3.015	**2**	**1**
.1	3.017	3.018	3.020	3.022	3.023	3.025	3.027	3.028	3.030	3.032	0	0
.2	3.033	3.035	3.036	3.038	3.040	3.041	3.043	3.045	3.046	3.048	0	0
.3	3.050	3.051	3.053	3.055	3.056	3.058	3.059	3.061	3.063	3.064	1	0
.4	3.066	3.068	3.069	3.071	3.072	3.074	3.076	3.077	3.079	3.081	1	0
9.5	3.082	3.084	3.085	3.087	3.089	3.090	3.092	3.094	3.095	3.097	1	1
.6	3.098	**3.100**	3.102	3.103	3.105	3.106	3.108	3.110	3.111	3.113	1	1
.7	3.114	3.116	3.118	3.119	3.121	3.122	3.124	3.126	3.127	3.129	1	1
.8	3.130	3.132	3.134	3.135	3.137	3.138	3.140	3.142	3.143	3.145	2	1
.9	3.146	3.148	3.150	3.151	3.153	3.154	3.156	3.158	3.159	3.161	2	1

SQUARE ROOTS AND SQUARES.

No.	0	1	2	3	4	5	6	7	8	9	Interpola. for Hundredths.	
10.	3.162	3.178	3.194	3.209	3.225	3.240	3.256	3.271	3.286	3.302	**16**	**14**
11.	3.317	3.332	3.347	3.362	3.376	3.391	3.406	3.421	3.435	3.450	2	1
12.	3.464	3.479	3.493	**3.507**	3.521	3.536	3.550	3.564	3.578	3.592	3	3
13.	3.606	3.619	3.633	3.647	3.661	3.674	3.688	3.701	3.715	3.728	5	4
14.	3.742	3.755	3.768	3.782	3.795	3.808	3.821	3.834	3.847	3.860	6	6
15.	3.873	3.886	3.899	3.912	3.924	3.937	3.950	3.962	3.975	3.987	8	7
16.	**4.000**	4.012	4.025	4.037	4.050	4.062	4.074	4.087	4.099	4.111	10	8
17.	4.123	4.135	4.147	4.159	4.171	4.183	4.195	4.207	4.219	4.231	11	10
18.	4.243	4.254	4.266	4.278	4.290	4.301	4.313	4.324	4.336	4.347	13	11
19.	4.359	4.370	4.382	4.393	4.405	4.416	4.427	4.438	4.450	4.461	14	13
20.	4.472	4.483	4.494	**4.506**	4.517	4.528	4.539	4.550	4.561	4.572	**12**	**10**
21.	4.583	4.593	4.604	4.615	4.626	4.637	4.648	4.658	4.669	4.680	1	1
22.	4.690	4.701	4.712	4.722	4.733	4.743	4.754	4.764	4.775	4.785	2	2
23.	4.796	4.806	4.817	4.827	4.837	4.848	4.858	4.868	4.879	4.889	4	3
24.	4.899	4.909	4.919	4.930	4.940	4.950	4.960	4.970	4.980	4.990	5	4
25.	**5.000**	5.010	5.020	5.030	5.040	5.050	5.060	5.070	5.079	5.089	6	5
26.	5.099	5.109	5.119	5.128	5.138	5.148	5.158	5.167	5.177	5.187	7	6
27.	5.196	5.206	5.215	5.225	5.235	5.244	5.254	5.263	5.273	5.282	8	7
28.	5.292	5.301	5.310	5.320	5.329	5.339	5.348	5.357	5.367	5.376	10	8
29.	5.385	5.394	5.404	5.413	5.422	5.431	5.441	5.450	5.459	5.468	11	9
30.	5.477	5.486	5.495	**5.505**	5.514	5.523	5.532	5.541	5.550	5.559	**9**	**8**
31.	5.568	5.577	5.586	5.595	5.604	5.612	5.621	5.630	5.639	5.648	1	1
32.	5.657	5.666	5.675	5.683	5.692	5.701	5.710	5.718	5.727	5.736	2	2
33.	5.745	5.753	5.762	5.771	5.779	5.788	5.797	5.805	5.814	5.822	3	2
34.	5.831	5.840	5.848	5.857	5.865	5.874	5.882	5.891	5.899	5.908	4	3
35.	5.916	5.925	5.933	5.941	5.950	5.958	5.967	5.975	5.983	5.992	5	4
36.	**6.000**	6.008	6.017	6.025	6.033	6.042	6.050	6.058	6.066	6.075	5	5
37.	6.083	6.091	6.099	6.107	6.116	6.124	6.132	6.140	6.148	6.156	6	6
38.	6.164	6.173	6.181	6.189	6.197	6.205	6.213	6.221	6.229	6.237	7	6
39.	6.245	6.253	6.261	6.269	6.277	6.285	6.293	6.301	6.309	6.317	8	7
40.	6.325	6.332	6.340	6.348	6.356	6.364	6.372	6.380	6.387	6.395	**8**	7
41.	6.403	6.411	6.419	6.427	6.434	6.442	6.450	6.458	6.465	6.473	1	1
42.	6.481	6.488	6.496	**6.504**	6.512	6.519	6.527	6.535	6.542	6.550	2	1
43.	6.557	6.565	6.573	6.580	6.588	6.595	6.603	6.611	6.618	6.626	2	2
44.	6.633	6.641	6.648	6.656	6.663	6.671	6.678	6.686	6.693	6.701	3	3
45.	6.708	6.716	6.723	6.731	6.738	6.745	6.753	6.760	6.768	6.775	4	4
46.	6.782	6.790	6.797	6.804	6.812	6.819	6.826	6.834	6.841	6.848	5	4
47.	6.856	6.863	6.870	6.877	6.885	6.892	6.899	6.907	6.914	6.921	6	5
48.	6.928	6.935	6.943	6.950	6.957	6.964	6.971	6.979	6.986	6.993	6	6
49.	**7.000**	7.007	7.014	7.021	7.029	7.036	7.043	7.050	7.057	7.064	7	6

SQUARE ROOTS AND SQUARES.

No.	0	1	2	3	4	5	6	7	8	9	Interpola. for Hundredths.	
50.	7.071	7.078	7.085	7.092	7.099	7.106	7.113	7.120	7.127	7.134	**7**	**6**
51.	7.141	7.148	7.155	7.162	7.169	7.176	7.183	7.190	7.197	7.204	1	1
52.	7.211	7.218	7.225	7.232	7.239	7.246	7.253	7.259	7.266	7.273	1	1
53.	7.280	7.287	7.294	7.301	7.308	7.314	7.321	7.328	7.335	7.342	2	2
54.	7.348	7.355	7.362	7.369	7.376	7.382	7.389	7.396	7.403	7.409	3	2
55.	7.416	7.423	7.430	7.436	7.443	7.450	7.457	7.463	7.470	7.477	4	3
56.	7.483	7.490	7.497	**7.503**	7.510	7.517	7.523	7.530	7.537	7.543	4	4
57.	7.550	7.556	7.563	7.570	7.576	7.583	7.589	7.596	7.603	7.609	5	4
58.	7.616	7.622	7.629	7.635	7.642	7.649	7.655	7.662	7.668	7.675	6	5
59.	7.681	7.688	7.694	7.701	7.707	7.714	7.720	7.727	7.733	7.740	6	5
60.	7.746	7.752	7.759	7.765	7.772	7.778	7.785	7.791	7.797	7.804	**7**	**6**
61.	7.810	7.817	7.823	7.829	7.836	7.842	7.849	7.855	7.861	7.868	1	1
62.	7.874	7.880	7.887	7.893	7.899	7.906	7.912	7.918	7.925	7.931	1	1
63.	7.937	7.944	7.950	7.956	7.962	7.969	7.975	7.981	7.987	7.994	2	2
64.	**8.000**	8.006	8.012	8.019	8.025	8.031	8.037	8.044	8.050	8.056	3	2
65.	8.062	8.068	8.075	8.081	8.087	8.093	8.099	8.106	8.112	8.118	4	3
66.	8.124	8.130	8.136	8.142	8.149	8.155	8.161	8.167	8.173	8.179	4	4
67.	8.185	8.191	8.198	8.204	8.210	8.216	8.222	8.228	8.234	8.240	5	4
68.	8.246	8.252	8.258	8.264	8.270	8.276	8.283	8.289	8.295	8.301	6	5
69.	8.307	8.313	8.319	8.325	8.331	8.337	8.343	8.349	8.355	8.361	6	5
70.	8.367	8.373	8.379	8.385	8.390	8.396	8.402	8.408	8.414	8.420	**6**	**5**
71.	8.426	8.432	8.438	8.444	8.450	8.456	8.462	8.468	8.473	8.479	1	1
72.	8.485	8.491	8.497	**8.503**	8.509	8.515	8.521	8.526	8.532	8.538	1	1
73.	8.544	8.550	8.556	8.562	8.567	8.573	8.579	8.585	8.591	8.597	2	2
74.	8.602	8.608	8.614	8.620	8.626	8.631	8.637	8.643	8.649	8.654	2	2
75.	8.660	8.666	8.672	8.678	8.683	8.689	8.695	8.701	8.706	8.712	3	3
76.	8.718	8.724	8.729	8.735	8.741	8.746	8.752	8.758	8.764	8.769	4	3
77.	8.775	8.781	8.786	8.792	8.798	8.803	8.809	8.815	8.820	8.826	4	4
78.	8.832	8.837	8.843	8.849	8.854	8.860	8.866	8.871	8.877	8.883	5	4
79.	8.888	8.894	8.899	8.905	8.911	8.916	8.922	8.927	8.933	8.939	5	5
80.	8.944	8.950	8.955	8.961	8.967	8.972	8.978	8.983	8.989	8.994	**6**	**5**
81.	**9.000**	9.006	9.011	9.017	9.022	9.028	9.033	9.039	9.044	9.050	1	1
82.	9.055	9.061	9.066	9.072	9.077	9.083	9.088	9.094	9.099	9.105	1	1
83.	9.110	9.116	9.121	9.127	9.132	9.138	9.143	9.149	9.154	9.160	2	2
84.	9.165	9.171	9.176	9.182	9.187	9.192	9.198	9.203	9.209	9.214	2	2
85.	9.220	9.225	9.230	9.236	9.241	9.247	9.252	9.257	9.263	9.268	3	3
86.	9.274	9.279	9.284	9.290	9.295	9.301	9.306	9.311	9.317	9.322	4	3
87.	9.327	9.333	9.338	9.343	9.349	9.354	9.359	9.365	9.370	9.375	4	4
88.	9.381	9.386	9.391	9.397	9.402	9.407	9.413	9.418	9.423	9.429	5	4
89.	9.434	9.439	9.445	9.450	9.455	9.460	9.466	9.471	9.476	9.482	5	5
90.	9.487	9.492	9.497	**9.503**	9.508	9.513	9.518	9.524	9.529	9.534	**5**	
91.	9.539	9.545	9.550	9.555	9.560	9.566	9.571	9.576	9.581	9.586	1	
92.	9.592	9.597	9.602	9.607	9.612	9.618	9.623	9.628	9.633	9.638	1	
93.	9.644	9.649	9.654	9.659	9.664	9.670	9.675	9.680	9.685	9.690	2	
94.	9.695	9.701	9.706	9.711	9.716	9.721	9.726	9.731	9.737	9.742	2	
95.	9.747	9.752	9.757	9.762	9.767	9.772	9.778	9.783	9.788	9.793	3	
96.	9.798	9.803	9.808	9.813	9.818	9.823	9.829	9.834	9.839	9.844	3	
97.	9.849	9.854	9.859	9.864	9.869	9.874	9.879	9.884	9.889	9.894	4	
98.	9.899	9.905	9.910	9.915	9.920	9.925	9.930	9.935	9.940	9.945	4	
99.	9.950	9.955	9.960	9.965	9.970	9.975	9.980	9.985	9.990	9.995	5	

RECIPROCALS.

No.	0	1	2	3	4	5	6	7	8	9	INTERPOLATION TABLES.			
1.00		0.9990	0.9980	0.9970	0.9960	0.9950	0.9940	0.9930	0.9921	0.9911	**-85**	**-75**	**-65**	**-55**
.01	0.9901	9891	9881	9872	9862	9852	9843	9833	9823	9814	9	8	7	6
.02	9804	9794	9785	9775	9766	9756	9747	9737	9728	9718	17	15	13	11
.03	9709	9699	9690	9680	9671	9662	9653	9643	9634	9625	26	23	20	17
.04	9615	9606	9597	9588	9579	9569	9560	9551	9542	9533	34	30	26	22
1.05	0.9524	0.9515	0.9506	0.9497	0.9488	0.9479	0.9470	0.9461	0.9452	0.9443	43	38	33	28
.06	9434	9425	9416	9407	9399	9390	9381	9372	9363	9355	51	45	39	33
.07	9346	9337	9328	9320	9311	9302	9294	9285	9276	9268	60	53	46	39
.08	9259	9251	9242	9234	9225	9217	9208	9200	9191	9183	68	60	52	44
.09	9174	9166	9158	9149	9141	9132	9124	9116	9107	9099	77	68	59	50
1.0		0.9901	0.9804	0.9709	0.9615	0.9524	0.9434	0.9346	0.9259	0.9174	**-45**	**-35**	**-25**	**-22**
.1	0.9091	9009	8929	8850	8772	8696	8621	8547	8475	8403	5	4	3	2
.2	8333	8264	8197	8130	8065	8000	7937	7874	7813	7752	9	7	5	4
.3	7692	7634	7576	7519	7463	7407	7353	7299	7246	7194	14	11	8	7
.4	7143	7092	7042	6993	6944	6897	6849	6803	6757	6711	18	14	10	9
1.5	0.6667	0.6623	0.6579	0.6536	0.6494	0.6452	0.6410	0.6369	0.6329	0.6289	23	18	13	11
.6	6250	6211	6173	6135	6098	6061	6024	5988	5952	5917	27	21	15	13
.7	5882	5848	5814	5780	5747	5714	5682	5650	5618	5587	32	25	18	15
.8	5556	5525	5495	5464	5435	5405	5376	5348	5319	5291	36	28	20	18
.9	5263	5236	5208	5181	5155	5128	5102	5076	5051	5025	41	32	23	20
2.0	0.5000	0.4975	0.4950	0.4926	0.4902	0.4878	0.4854	0.4831	0.4808	0.4785	**-18**	**-16**	**-14**	**-12**
.1	4762	4739	4717	4695	4673	4651	4630	4608	4587	4566	2	2	1	1
.2	4545	4525	4505	4484	4464	4444	4425	4405	4386	4367	4	3	3	2
.3	4348	4329	4310	4292	4274	4255	4237	4219	4202	4184	5	5	4	4
.4	4167	4149	4132	4115	4098	4082	4065	4049	4032	4016	7	6	6	5
2.5	0.4000	0.3984	0.3968	0.3953	0.3937	0.3922	0.3906	0.3891	0.3876	0.3861	9	8	7	6
.6	3846	3831	3817	3802	3788	3774	3759	3745	3731	3717	11	10	8	7
.7	3704	3690	3676	3663	3650	3636	3623	3610	3597	3584	13	11	10	8
.8	3571	3559	3546	3534	3521	3509	3497	3484	3472	3460	14	13	11	10
.9	3448	3436	3425	3413	3401	3390	3378	3367	3356	3344	16	14	13	11
3.0	0.3333	0.3322	0.3311	0.3300	0.3289	0.3279	0.3268	0.3257	0.3247	0.3236	**-11**	**-9**	**-8**	**-7**
.1	3226	3215	3205	3195	3185	3175	3165	3155	3145	3135	1	1	1	1
.2	3125	3115	3106	3096	3086	3077	3067	3058	3049	3040	2	2	2	1
.3	3030	3021	3012	3003	2994	2985	2976	2967	2959	2950	3	3	2	2
.4	2941	2933	2924	2915	2907	2899	2890	2882	2874	2865	4	4	3	3
3.5	0.2857	0.2849	0.2841	0.2833	0.2825	0.2817	0.2809	0.2801	0.2793	0.2786	6	5	4	4
.6	2778	2770	2762	2755	2747	2740	2732	2725	2717	2710	7	5	5	4
.7	2703	2695	2688	2681	2674	2667	2660	2653	2646	2639	8	6	6	5
.8	2632	2625	2618	2611	2604	2597	2591	2584	2577	2571	9	7	6	6
.9	2564	2558	2551	2545	2538	2532	2525	2519	2513	2506	10	8	7	6
4.0	0.2500	0.2494	0.2488	0.2481	0.2475	0.2469	0.2463	0.2457	0.2451	0.2445	**-6**	**-5**	**-4**	
.1	2439	2433	2427	2421	2415	2410	2404	2398	2392	2387	1	1	0	
.2	2381	2375	2370	2364	2358	2353	2347	2342	2336	2331	1	1	1	
.3	2326	2320	2315	2309	2304	2299	2294	2288	2283	2278	2	2	1	
.4	2273	2268	2262	2257	2252	2247	2242	2237	2232	2227	2	2	2	
4.5	0.2222	0.2217	0.2212	0.2208	0.2203	0.2198	0.2193	0.2188	0.2183	0.2179	3	3	2	
.6	2174	2169	2165	2160	2155	2151	2146	2141	2137	2132	4	3	2	
.7	2128	2123	2119	2114	2110	2105	2101	2096	2092	2088	4	4	3	
.8	2083	2079	2075	2070	2066	2062	2058	2053	2049	2045	5	4	3	
.9	2041	2037	2033	2028	2024	2020	2016	2012	2008	2004	5	5	4	

RECIPROCALS.

No.	0	1	2	3	4	5	6	7	8	9	Interpolation for thous.	
5.0	0.2000	0.1996	0.1992	0.1988	0.1984	0.1980	0.1976	0.1972	0.1969	0.1965	-4	-3
.1	1961	1957	1953	1949	1946	1942	1938	1934	1931	1927	0	0
.2	1923	1919	1916	1912	1908	1905	1901	1898	1894	1890	1	1
.3	1887	1883	1880	1876	1873	1869	1866	1862	1859	1855	1	1
.4	1852	1848	1845	1842	1838	1835	1832	1828	1825	1821	2	1
5.5	0.1818	0.1815	0.1812	0.1808	0.1805	0.1802	0.1799	0.1795	0.1792	0.1789	2	2
.6	1786	1783	1779	1776	1773	1770	1767	1764	1761	1757	2	2
.7	1754	1751	1748	1745	1742	1739	1736	1733	1730	1727	3	2
.8	1724	1721	1718	1715	1712	1709	1706	1704	1701	1698	3	2
.9	1695	1692	1689	1686	1684	1681	1678	1675	1672	1669	4	3
6.0	0.1667	0.1664	0.1661	0.1658	0.1656	0.1653	0.1650	0.1647	0.1645	0.1642	-3	-2
.1	1639	1637	1634	1631	1629	1626	1623	1621	1618	1616	0	0
.2	1613	1610	1608	1605	1603	1600	1597	1595	1592	1590	1	0
.3	1587	1585	1582	1580	1577	1575	1572	1570	1567	1565	1	1
.4	1563	1560	1558	1555	1553	1550	1548	1546	1543	1541	1	1
6.5	0.1538	0.1536	0.1534	0.1531	0.1529	0.1527	0.1524	0.1522	0.1520	0.1517	2	1
.6	1515	1513	1511	1508	1506	1504	1502	1499	1497	1495	2	1
.7	1493	1490	1488	1486	1484	1481	1479	1477	1475	1473	2	1
.8	1471	1468	1466	1464	1462	1460	1458	1456	1453	1451	2	2
.9	1449	1447	1445	1443	1441	1439	1437	1435	1433	1431	3	2
7.0	0.1429	0.1427	0.1425	0.1422	0.1420	0.1418	0.1416	0.1414	0.1412	0.1410	-2	-1
.1	1408	1406	1404	1403	1401	1399	1397	1395	1393	1391	0	0
.2	1389	1387	1385	1383	1381	1379	1377	1376	1374	1372	0	0
.3	1370	1368	1366	1364	1362	1361	1359	1357	1355	1353	1	0
.4	1351	1350	1348	1346	1344	1342	1340	1339	1337	1335	1	0
7.5	0.1333	0.1332	0.1330	0.1328	0.1326	0.1325	0.1323	0.1321	0.1319	0.1318	1	1
.6	1316	1314	1312	1311	1309	1307	1305	1304	1302	1300	1	1
.7	1299	1297	1295	1294	1292	1290	1289	1287	1285	1284	1	1
.8	1282	1280	1279	1277	1276	1274	1272	1271	1269	1267	2	1
.9	1266	1264	1263	1261	1259	1258	1256	1255	1253	1252	2	1
8.0	0.1250	0.1248	0.1247	0.1245	0.1244	0.1242	0.1241	0.1239	0.1238	0.1236	-2	-1
.1	1235	1233	1232	1230	1229	1227	1225	1224	1222	1221	0	0
.2	1220	1218	1217	1215	1214	1212	1211	1209	1208	1206	0	0
.3	1205	1203	1202	1200	1199	1198	1196	1195	1193	1192	1	0
.4	1190	1189	1188	1186	1185	1183	1182	1181	1179	1178	1	0
8.5	0.1176	0.1175	0.1174	0.1172	0.1171	0.1170	0.1168	0.1167	0.1166	0.1164	1	1
.6	1163	1161	1160	1159	1157	1156	1155	1153	1152	1151	1	1
.7	1149	1148	1147	1145	1144	1143	1142	1140	1139	1138	1	1
.8	1136	1135	1134	1133	1131	1130	1129	1127	1126	1125	2	1
.9	1124	1122	1121	1120	1119	1117	1116	1115	1114	1112	2	1
9.0	0.1111	0.1110	0.1109	0.1107	0.1106	0.1105	0.1104	0.1103	0.1101	0.1100	-2	-1
.1	1099	1098	1096	1095	1094	1093	1092	1091	1089	1088	0	0
.2	1087	1086	1085	1083	1082	1081	1080	1079	1078	1076	0	0
.3	1075	1074	1073	1072	1071	1070	1068	1067	1066	1065	1	0
.4	1064	1063	1062	1060	1059	1058	1057	1056	1055	1054	1	0
9.5	0.1053	0.1052	0.1050	0.1049	0.1048	0.1047	0.1046	0.1045	0.1044	0.1043	1	1
.6	1042	1041	1040	1038	1037	1036	1035	1034	1033	1032	1	1
.7	1031	1030	1029	1028	1027	1026	1025	1024	1022	1021	1	1
.8	1020	1019	1018	1017	1016	1015	1014	1013	1012	1011	2	1
.9	1010	1009	1008	1007	1006	1005	1004	1003	1002	1001	2	1

SLIDE-WIRE RATIOS.

cm.	$0^{mm.}$	$1^{mm.}$	$2^{mm.}$	$3^{mm.}$	$4^{mm.}$	$5^{mm.}$	$6^{mm.}$	$7^{mm.}$	$8^{mm.}$	$9^{mm.}$
0	0.0000	0.0010	0.0020	0.0030	0.0040	0.0050	0.0060	0.0071	0.0081	0.0091
1	0101	0111	0122	0132	0142	0152	0163	0173	0183	0194
2	0204	0215	0225	0235	0246	0256	0267	0278	0288	0299
3	0309	0320	0331	0341	0352	0363	0373	0384	0395	0406
4	0417	0428	0438	0449	0460	0471	0482	0493	0504	0515
5	0.0526	0.0537	0.0549	0.0560	0.0571	0.0582	0.0593	0.0605	0.0616	0.0627
6	0638	0650	0661	0672	0684	0695	0707	0718	0730	0741
7	0753	0764	0776	0788	0799	0811	0823	0834	0846	0858
8	0870	0881	0893	0905	0917	0929	0941	0953	0965	0977
9	0989	1001	1013	1025	1038	1050	1062	1074	1087	1099
10	0.1111	0.1124	0.1136	0.1148	0.1161	0.1173	0.1186	0.1198	0.1211	0.1223
11	1236	1249	1261	1274	1287	1299	1312	1325	1338	1351
12	1364	1377	1390	1403	1416	1429	1442	1455	1468	1481
13	1494	1508	1521	1534	1547	1561	1574	1588	1601	1614
14	1628	1641	1655	1669	1682	1696	1710	1723	1737	1751
15	0.1765	0.1779	0.1793	0.1806	0.1820	0.1834	0.1848	0.1862	0.1877	0.1891
16	1905	1919	1933	1947	1962	1976	1990	2005	2019	2034
17	2048	2063	2077	2092	2107	2121	2136	2151	2166	2180
18	2195	2210	2225	2240	2255	2270	2285	2300	2315	2331
19	2346	2361	2376	2392	2407	2422	2438	2453	2469	2484
20	0.2500	0.2516	0.2531	0.2547	0.2563	0.2579	0.2595	0.2610	0.2626	0.2642
21	2658	2674	2690	2707	2723	2739	2755	2771	2788	2804
22	2821	2837	2854	2870	2887	2903	2920	2937	2953	2970
23	2987	3004	3021	3038	3055	3072	3089	3106	3123	3141
24	3158	3175	3193	3210	3228	3245	3263	3280	3298	3316
25	0.3333	0.3351	0.3369	0.3387	0.3405	0.3423	0.3441	0.3459	0.3477	0.3495
26	3514	3532	3550	3569	3587	3605	3624	3643	3661	3680
27	3699	3717	3736	3755	3774	3793	3812	3831	3850	3870
28	3889	3908	3928	3947	3967	3986	4006	4025	4045	4065
29	4085	4104	4124	4144	4164	4184	4205	4225	4245	4265
30	0.4286	0.4306	0.4327	0.4347	0.4368	0.4389	0.4409	0.4430	0.4451	0.4472
31	4493	4514	4535	4556	4577	4599	4620	4641	4663	4684
32	4706	4728	4749	4771	4793	4815	4837	4859	4881	4903
33	4925	4948	4970	4993	5015	5038	5060	5083	5106	5129
34	5152	5175	5198	5221	5244	5267	5291	5314	5337	5361
35	0.5385	0.5408	0.5432	0.5456	0.5480	0.5504	0.5528	0.5552	0.5576	0.5601
36	5625	5650	5674	5699	5723	5748	5773	5798	5823	5848
37	5873	5898	5924	5949	5974	6000	6026	6051	6077	6103
38	6129	6155	6181	6208	6234	6260	6287	6313	6340	6367
39	6393	6420	6447	6475	6502	6529	6556	6584	6611	6639
40	0.6667	0.6695	0.6722	0.6750	0.6779	0.6807	0.6835	0.6863	0.6892	0.6921
41	6949	6978	7007	7036	7065	7094	7123	7153	7182	7212
42	7241	7271	7301	7331	7361	7391	7422	7452	7483	7513
43	7544	7575	7606	7637	7668	7699	7731	7762	7794	7825
44	7857	7889	7921	7953	7986	8018	8051	8083	8116	8149
45	0.8182	0.8215	0.8248	0.8282	0.8315	0.8349	0.8382	0.8416	0.8450	0.8484
46	8519	8553	8587	8622	8657	8692	8727	8762	8797	8832
47	8868	8904	8939	8975	9011	9048	9084	9121	9157	9194
48	9231	9268	9305	9342	9380	9418	9455	9493	9531	9570
49	9608	9646	9685	9724	9763	9802	9841	9881	9920	9960

SLIDE-WIRE RATIOS.

cm.	0mm.	1mm.	2mm.	3mm.	4mm.	5mm.	6mm.	7mm.	8mm.	9mm.
50	1.000	1.004	1.008	1.012	1.016	1.020	1.024	1.028	1.033	1.037
51	1.041	1.045	1.049	1.053	1.058	1.062	1.066	1.070	1.075	1.079
52	1.083	1.088	1.092	1.096	1.101	1.105	1.110	1.114	1.119	1.123
53	1.128	1.132	1.137	1.141	1.146	1.151	1.155	1.160	1.165	1.169
54	1.174	1.179	1.183	1.188	1.193	1.198	1.203	1.208	1.212	1.217
55	1.222	1.227	1.232	1.237	1.242	1.247	1.252	1.257	1.262	1.268
56	1.273	1.278	1.283	1.288	1.294	1.299	1.304	1.309	1.315	1.320
57	1.326	1.331	1.336	1.342	1.347	1.353	1.358	1.364	1.370	1.375
58	1.381	1.387	1.392	1.398	1.404	1.410	1.415	1.421	1.427	1.433
59	1.439	1.445	1.451	1.457	1.463	1.469	1.475	1.481	1.488	1.494
60	1.500	1.506	1.513	1.519	1.525	1.532	1.538	1.545	1.551	1.558
61	1.564	1.571	1.577	1.584	1.591	1.597	1.604	1.611	1.618	1.625
62	1.632	1.639	1.646	1.653	1.660	1.667	1.674	1.681	1.688	1.695
63	1.703	1.710	1.717	1.725	1.732	1.740	1.747	1.755	1.762	1.770
64	1.778	1.786	1.793	1.801	1.809	1.817	1.825	1.833	1.841	1.849
65	1.857	1.865	1.874	1.882	1.890	1.899	1.907	1.915	1.924	1.933
66	1.941	1.950	1.959	1.967	1.976	1.985	1.994	2.003	2.012	2.021
67	2.030	2.040	2.049	2.058	2.067	2.077	2.086	2.096	2.106	2.115
68	2.125	2.135	2.145	2.155	2.165	2.175	2.185	2.195	2.205	2.215
69	2.226	2.236	2.247	2.257	2.268	2.279	2.289	2.300	2.311	2.322
70	2.333	2.344	2.356	2.367	2.378	2.390	2.401	2.413	2.425	2.436
71	2.448	2.460	2.472	2.484	2.497	2.509	2.521	2.534	2.546	2.559
72	2.571	2.584	2.597	2.610	2.623	2.636	2.650	2.663	2.676	2.690
73	2.704	2.717	2.731	2.745	2.759	2.774	2.788	2.802	2.817	2.831
74	2.846	2.861	2.876	2.891	2.906	2.922	2.937	2.953	2.968	2.984
75	3.000	3.016	3.032	3.049	3.065	3.082	3.098	3.115	3.132	3.149
76	3.167	3.184	3.202	3.219	3.237	3.255	3.274	3.292	3.310	3.329
77	3.348	3.367	3.386	3.405	3.425	3.444	3.464	3.484	3.505	3.525
78	3.545	3.566	3.587	3.608	3.630	3.651	3.673	3.695	3.717	3.739
79	3.762	3.785	3.808	3.831	3.854	3.878	3.902	3.926	3.950	3.975
80	4.000	4.025	4.051	4.076	4.102	4.128	4.155	4.181	4.208	4.236
81	4.263	4.291	4.319	4.348	4.376	4.405	4.435	4.465	4.495	4.525
82	4.556	4.587	4.618	4.650	4.682	4.714	4.747	4.780	4.814	4.848
83	4.882	4.917	4.952	4.988	5.024	5.061	5.098	5.135	5.173	5.211
84	5.250	5.289	5.329	5.369	5.410	5.452	5.494	5.536	5.579	5.623
85	5.667	5.711	5.757	5.803	5.849	5.897	5.944	5.993	6.042	6.092
86	6.143	6.194	6.246	6.299	6.353	6.407	6.463	6.519	6.576	6.634
87	6.692	6.752	6.813	6.874	6.937	7.000	7.065	7.130	7.197	7.264
88	7.333	7.403	7.475	7.547	7.621	7.696	7.772	7.850	7.929	8.009
89	8.091	8.174	8.259	8.346	8.434	8.524	8.615	8.709	8.804	8.901
90	9.000	9.101	9.204	9.309	9.417	9.526	9.638	9.753	9.870	9.989
91	10.11	10.33	10.36	10.49	10.63	10.77	10.90	11.05	11.20	11.35
92	11.50	11.66	11.82	11.99	12.16	12.33	12.51	12.70	12.89	13.08
93	13.29	13.49	13.71	13.93	14.15	14.38	14.63	14.87	15.13	15.39
94	15.67	15.95	16.24	16.54	16.86	17.18	17.52	17.87	18.23	18.61
95	19.00	19.41	19.83	20.28	20.74	21.22	21.73	22.26	22.81	23.39
96	24.00	24.64	25.32	26.03	26.78	27.57	28.41	29.30	30.25	31.26
97	32.33	33.48	34.71	36.04	37.46	39.00	40.67	42.48	44.45	46.62
98	49.00	51.6	54.6	57.8	61.5	65.7	70.4	75.9	82.3	89.9
99	99.0	110.	124.	142.	166.	199.	249.	332.	499.	999.

NATURAL SINES AND COSINES

TO

FOUR PLACES.

Note. For cosines use right-hand column of degrees and lower line of tenths.

Deg.	°.0	°.1	°.2	°.3	°.4	°.5	°.6	°.7	°.8	°.9		Interpola for h'dths	
0°	0.0000	0.0017	0.0035	0.0052	0.0070	0.0087	0.0105	0.0122	0.0140	0.0157	89	18	17
1	0175	0192	0209	0227	0244	0262	0279	0297	0314	0332	88	2	2
2	0349	0366	0384	0401	0419	0436	0454	0471	0488	0506	87	4	3
3	0523	0541	0558	0576	0593	0610	0628	0645	0663	.0680	86	5	5
4	0698	0715	0732	0750	0767	0785	0802	0819	0837	0854	85	7	7
5	0.0872	0.0889	0.0906	0.0924	0.0941	0.0958	0.0976	0.0993	0.1011	0.1028	84	9	9
6	1045	1063	1080	1097	1115	1132	1149	1167	1184	1201	83	11	10
7	1219	1236	1253	1271	1288	1305	1323	1340	1357	1374	82	13	12
8	1392	1409	1426	1444	1461	1478	1495	1513	1530	1547	81	14	14
9	1564	1582	1599	1616	1633	1650	1668	1685	1702	1719	80°	16	15
10°	0.1736	0.1754	0.1771	0.1788	0.1805	0.1822	0.1840	0.1857	0.1874	0.1891	79	17	16
11	1908	1925	1942	1959	1977	1994	2011	2028	2045	2062	78	2	2
12	2079	2096	2113	2130	2147	2164	2181	2198	2215	2232	77	3	3
13	2250	2267	2284	2300	2317	2334	2351	2368	2385	2402	76	5	5
14	2419	2436	2453	2470	2487	2504	2521	2538	2554	2571	75	7	6
15	0.2588	0.2605	0.2622	0.2639	0.2656	0.2672	0.2689	0.2706	0.2723	0.2740	74	9	8
16	2756	2773	2790	2807	2823	2840	2857	2874	2890	2907	73	10	10
17	2924	2940	2957	2974	2990	3007	3024	3040	3057	3074	72	12	11
18	3090	3107	3123	3140	3156	3173	3190	3206	3223	3239	71	14	13
19	3256	3272	3289	3305	3322	3338	3355	3371	3387	3404	70°	15	14
20°	0.3420	0.3437	0.3453	0.3469	0.3486	0.3502	0.3518	0.3535	0.3551	0.3567	69	16	15
21	3584	3600	3616	3633	3649	3665	3681	3697	3714	3730	68	2	2
22	3746	3762	3778	3795	3811	3827	3843	3859	3875	3891	67	3	3
23	3907	3923	3939	3955	3971	3987	4003	4019	4035	4051	66	5	5
24	4067	4083	4099	4115	4131	4147	4163	4179	4195	4210	65	6	6
25	0.4226	0.4242	0.4258	0.4274	0.4289	0.4305	0.4321	0.4337	0.4352	0.4368	64	8	8
26	4384	4399	4415	4431	4446	4462	4478	4493	4509	4524	63	10	9
27	4540	4555	4571	4586	4602	4617	4633	4648	4664	4679	62	11	11
28	4695	4710	4726	4741	4756	4772	4787	4802	4818	4833	61	13	12
29	4848	4863	4879	4894	4909	4924	4939	4955	4970	4985	60°	14	14
30°	0.5000	0.5015	0.5030	0.5045	0.5060	0.5075	0.5090	0.5105	0.5120	0.5135	59	14	13
31	5150	5165	5180	5195	5210	5225	5240	5255	5270	5284	58	1	1
32	5299	5314	5329	5344	5358	5373	5388	5402	5417	5432	57	3	3
33	5446	5461	5476	5490	5505	5519	5534	5548	5563	5577	56	4	4
34	5592	5606	5621	5635	5650	5664	5678	5693	5707	5721	55	6	5
35	0.5736	0.5750	0.5764	0.5779	0.5793	0.5807	0.5821	0.5835	0.5850	0.5864	54	7	7
36	5878	5892	5906	5920	5934	5948	5962	5976	5990	6004	53	8	8
37	6018	6032	6046	6060	6074	6088	6101	6115	6129	6143	52	10	9
38	6157	6170	6184	6198	6211	6225	6239	6252	6266	6280	51	11	10
39	6293	6307	6320	6334	6347	6361	6374	6388	6401	6414	50°	13	12
	1°.0	°.9	°.8	°.7	°.6	°.5	°.4	°.3	°.2	°.1	Deg.	Interpola. for h'dths	

NAT. COS. 4 PL.

NATURAL SINES AND COSINES.

4 PL. NAT. SIN.

Deg.	°.0	°.1	°.2	°.3	°.4	°.5	°.6	°.7	°.8	°.9		Interpola. for h'dths	
40°	0.6428	0.6441	0.6455	0.6468	0.6481	0.6494	0.6508	0.6521	0.6534	0.6547	49	**13**	**12**
41	6561	6574	6587	6600	6613	6626	6639	6652	6665	6678	48	1	1
42	6691	6704	6717	6730	6743	6756	6769	6782	6794	6807	47	3	2
43	6820	6833	6845	6858	6871	6884	6896	6909	6921	6934	46	4	4
44	6947	6959	6972	6984	6997	7009	7022	7034	7046	7059	45	5	5
45	0.7071	0.7083	0.7096	0.7108	0.7120	0.7133	0.7145	0.7157	0.7169	0.7181	44	7	6
46	7193	7206	7218	7230	7242	7254	7266	7278	7290	7302	43	8	7
47	7314	7325	7337	7349	7361	7373	7385	7396	7408	7420	42	9	8
48	7431	7443	7455	7466	7478	7490	7501	7513	7524	7536	41	10	10
49	7547	7559	7570	7581	7593	7604	7615	7627	7638	7649	40°	12	11
50°	0.7660	0.7672	0.7683	0.7694	0.7705	0.7716	0.7727	0.7738	0.7749	0.7760	39	**11**	**9**
51	7771	7782	7793	7804	7815	7826	7837	7848	7859	7869	38	1	1
52	7880	7891	7902	7912	7923	7934	7944	7955	7965	7976	37	2	2
53	7986	7997	8007	8018	8028	8039	8049	8059	8070	8080	36	3	3
54	8090	8100	8111	8121	8131	8141	8151	8161	8171	8181	35	4	4
55	0.8192	0.8202	0.8211	0.8221	0.8231	0.8241	0.8251	0.8261	0.8271	0.8281	34	6	5
56	8290	8300	8310	8320	8329	8339	8348	8358	8368	8377	33	7	5
57	8387	8396	8406	8415	8425	8434	8443	8453	8462	8471	32	8	6
58	8480	8490	8499	8508	8517	8526	8536	8545	8554	8563	31	9	7
59	8572	8581	8590	8599	8607	8616	8625	8634	8643	8652	30°	10	8
60°	0.8660	0.8669	0.8678	0.8686	0.8695	0.8704	0.8712	0.8721	0.8729	0.8738	29	**8**	**7**
61	8746	8755	8763	8771	8780	8788	8796	8805	8813	8821	28	1	1
62	8829	8838	8846	8854	8862	8870	8878	8886	8894	8902	27	2	1
63	8910	8918	8926	8934	8942	8949	8957	8965	8973	8980	26	2	2
64	8988	8996	9003	9011	9018	9026	9033	9041	9048	9056	25	3	3
65	0.9063	0.9070	0.9078	0.9085	0.9092	0.9100	0.9107	0.9114	0.9121	0.9128	24	4	4
66	9135	9143	9150	9157	9164	9171	9178	9184	9191	9198	23	5	4
67	9205	9212	9219	9225	9232	9239	9245	9252	9259	9265	22	6	5
68	9272	9278	9285	9291	9298	9304	9311	9317	9323	9330	21	6	6
69	9336	9342	9348	9354	9361	9367	9373	9379	9385	9391	20°	7	6
70°	0.9397	0.9403	0.9409	0.9415	0.9421	0.9426	0.9432	0.9438	0.9444	0.9449	19	**6**	**4**
71	9455	9461	9466	9472	9478	9483	9489	9494	9500	9505	18	1	0
72	9511	9516	9521	9527	9532	9537	9542	9548	9553	9558	17	1	1
73	9563	9568	9573	9578	9583	9588	9593	9598	9603	9608	16	2	1
74	9613	9617	9622	9627	9632	9636	9641	9646	9650	9655	15	2	2
75	0.9659	0.9664	0.9668	0.9673	0.9677	0.9681	0.9686	0.9690	0.9694	0.9699	14	3	2
76	9703	9707	9711	9715	9720	9724	9728	9732	9736	9740	13	4	2
77	9744	9748	9751	9755	9759	9763	9767	9770	9774	9778	12	4	3
78	9781	9785	9789	9792	9796	9799	9803	9806	9810	9813	11	5	3
79	9816	9820	9823	9826	9829	9833	9836	9839	9842	9845	10°	5	4
80°	0.9848	0.9851	0.9854	0.9857	0.9860	0.9863	0.9866	0.9869	0.9871	0.9874	9	**3**	**2**
81	9877	9880	9882	9885	9888	9890	9893	9895	9898	9900	8	0	0
82	9903	9905	9907	9910	9912	9914	9917	9919	9921	9923	7	1	0
83	9925	9928	9930	9932	9934	9936	9938	9940	9942	9943	6	1	1
84	9945	9947	9949	9951	9952	9954	9956	9957	9959	9960	5	1	1
85	0.9962	0.9963	0.9965	0.9966	0.9968	0.9969	0.9971	0.9972	0.9973	0.9974	4	2	1
86	9976	9977	9978	9979	9980	9981	9982	9983	9984	9985	3	2	1
87	9986	9987	9988	9989	9990	9990	9991	9992	9993	9993	2	2	1
88	9994	9995	9995	9996	9996	9997	9997	9997	9998	9998	1	2	2
89	9998	9999	9999	9999	9999	1.000	1.000	1.000	1.000	1.000	0°	3	2
	1°.0	°.9	°.8	°.7	°.6	°.5	°.4	°.3	°.2	°.1	Deg.	Interpola. for h'dths	

4 PL. NAT. COS.

NATURAL TANGENTS AND COTANGENTS

TO

FOUR PLACES.

Note. For cotangents use right-hand column of degrees and lower line of tenths

Deg.	°.0	°.1	°.2	°.3	°.4	°.5	°.6	°.7	°.8	°.9		Interpola. for h'dths	
0°	0.0000	0.0017	0.0035	0.0052	0.0070	0.0087	0.0105	0.0122	0.0140	0.0157	89	17	18
1	0175	0192	0209	0227	0244	0262	0279	0297	0314	0332	88	2	2
2	0349	0367	0384	0402	0419	0437	0454	0472	0489	0507	87	3	4
3	0524	0542	0559	0577	0594	0612	0629	0647	0664	0682	86	5	5
4	0699	0717	0734	0752	0769	0787	0805	0822	0840	0857	85	7	7
5	0.0875	0.0892	0.0910	0.0928	0.0945	0.0963	0.0981	0.0998	0.1016	0.1033	84	9	9
6	1051	1069	1086	1104	1122	1139	1157	1175	1192	1210	83	10	11
7	1228	1246	1263	1281	1299	1317	1334	1352	1370	1388	82	12	13
8	1405	1423	1441	1459	1477	1495	1512	1530	1548	1566	81	14	14
9	1584	1602	1620	1638	1655	1673	1691	1709	1727	1745	80°	15	16
10°	0.1763	0.1781	0.1799	0.1817	0.1835	0.1853	0.1871	0.1890	0.1908	0.1926	79	19	20
11	1944	[illegible]	1980	1998	2016	2035	2053	2071	2089	2107	78	2	2
12	2126	2144	2162	2180	2199	2217	2235	2254	2272	2290	77	4	4
13	2309	2327	2345	2364	2382	2401	2419	2438	2456	2475	76	6	6
14	2493	2512	2530	2549	2568	2586	2605	2623	2642	2661	75	8	8
15	0.2679	0.2698	0.2717	0.2736	0.2754	0.2773	0.2792	0.2811	0.2830	0.2849	74	10	10
16	2867	2886	2905	2924	2943	2962	2981	3000	3019	3038	73	11	12
17	3057	3076	3096	3115	3134	3153	3172	3191	3211	3230	72	13	14
18	3249	3269	3288	3307	3327	3346	3365	3385	3404	3424	71	15	16
19	3443	3463	3482	3502	3522	3541	3561	3581	3600	3620	70°	17	18
20°	0.3640	0.3659	0.3679	0.3699	0.3719	0.3739	0.3759	0.3779	0.3799	0.3819	69	22	24
21	3839	3859	3879	3899	3919	3939	3959	3979	4000	4020	68	2	2
22	4040	4061	4081	4101	4122	4142	4163	4183	4204	4224	67	4	5
23	4245	4265	4286	4307	4327	4348	4369	4390	4411	4431	66	7	7
24	4452	4473	4494	4515	4536	4557	4578	4599	4621	4642	65	9	10
25	0.4663	0.4684	0.4706	0.4727	0.4748	0.4770	0.4791	0.4813	0.4834	0.4856	64	11	12
26	4877	4899	4921	4942	4964	4986	5008	5029	5051	5073	63	13	14
27	5095	5117	5139	5161	5184	5206	5228	5250	5272	5295	62	15	17
28	5317	5340	5362	5384	5407	5430	5452	5475	5498	5520	61	18	19
29	5543	5566	5589	5612	5635	5658	5681	5704	5727	5750	60°	20	22
30°	0.5774	0.5797	0.5820	0.5844	0.5867	0.5890	0.5914	0.5938	0.5961	0.5985	59	26	28
31	6009	6032	6056	6080	6104	6128	6152	6176	6200	6224	58	3	3
32	6249	6273	6297	6322	6346	6371	6395	6420	6445	6469	57	6	6
33	6494	6519	6544	6569	6594	6619	6644	6669	6694	6720	56	8	8
34	6745	6771	6796	6822	6847	6873	6899	6924	6950	6976	55	10	11
35	0.7002	0.7028	0.7054	0.7080	0.7107	0.7133	0.7159	0.7186	0.7212	0.7239	54	13	14
36	7265	7292	7319	7346	7373	7400	7427	7454	7481	7508	53	16	17
37	7536	7563	7590	7618	7646	7673	7701	7729	7757	7785	52	18	20
38	7813	7841	7869	7898	7926	7954	7983	8012	8040	8069	51	21	22
39	8098	8127	8156	8185	8214	8243	8273	8302	8332	8361	50°	23	25
	1°.0	°.9	°.8	°.7	°.6	°.5	°.4	°.3	°.2	°.1	Deg.	Interpola. for h'dths	

NATURAL TANGENTS AND COTANGENTS.

4 PL. NAT. TAN.

Deg.	°.0	°.1	°.2	°.3	°.4	°.5	°.6	°.7	°.8	°.9		Interpola. for h'dths
40°	0.8391	0.8421	0.8451	0.8481	0.8511	0.8541	0.8571	0.8601	0.8632	0.8662	49	30 40
41	8693	8724	8754	8785	8816	8847	8878	8910	8941	8972	48	3 4
42	9004	9036	9067	9099	9131	9163	9195	9228	9260	9293	47	6 8
43	9325	9358	9391	9424	9457	9490	9523	9556	9590	9623	46	9 12
44	9657	9691	9725	9759	9793	9827	9861	9896	9930	9965	45	12 16
45	1.0000	1.0035	1.0070	1.0105	1.0141	1.0176	1.0212	1.0247	1.0283	1.0319	44	15 20
46	0355	0392	0428	0464	0501	0538	0575	0612	0649	0686	43	18 24
47	0724	0761	0799	0837	0875	0913	0951	0990	1028	1067	42	21 28
48	1106	1145	1184	1224	1263	1303	1343	1383	1423	1463	41	24 32
49	1504	1544	1585	1626	1667	1708	1750	1792	1833	1875	40°	27 36
50°	1.1918	1.1960	1.2002	1.2045	1.2088	1.2131	1.2174	1.2218	1.2261	1.2305	39	50 60
51	2349	2393	2437	2482	2527	2572	2617	2662	2708	2753	38	5 6
52	2799	2846	2892	2938	2985	3032	3079	3127	3175	3222	37	10 12
53	3270	3319	3367	3416	3465	3514	3564	3613	3663	3713	36	15 18
54	3764	3814	3865	3916	3968	4019	4071	4124	4176	4229	35	20 24
55	1.4281	1.4335	1.4388	1.4442	1.4496	1.4550	1.4605	1.4659	1.4715	1.4770	34	25 30
56	4826	4882	4938	4994	5051	5108	5166	5224	5282	5340	33	30 36
57	5399	5458	5517	5577	5637	5697	5757	5818	5880	5941	32	35 42
58	6003	6066	6128	6191	6255	6319	6383	6447	6512	6577	31	40 48
59	6643	6709	6775	6842	6909	6977	7045	7113	7182	7251	30°	45 54
60°	1.7321	1.7391	1.7461	1.7532	1.7603	1.7675	1.7747	1.7820	1.7893	1.7966	29	70 80
61	8040	8115	8190	8265	8341	8418	8495	8572	8650	8728	28	7 8
62	8807	8887	8967	9047	9128	9210	9292	9375	9458	9542	27	14 16
63	1.9626	1.9711	1.9797	1.9883	1.9970	2.0057	2.0145	2.0233	2.0323	2.0413	26	21 24
64	2.0503	2.0594	2.0686	2.0778	2.0872	2.0965	2.1060	2.1155	2.1251	2.1348	25	28 32
65	2.1445	2.1543	2.1642	2.1742	2.1842	2.1943	2.2045	2.2148	2.2251	2.2355	24	35 40
66	2460	2566	2673	2781	2889	2998	3109	3220	3332	3445	23	42 48
67	3559	3673	3789	3906	4023	4142	4262	4383	4504	4627	22	49 56
68	4751	4876	5002	5129	5257	5386	5517	5649	5782	5916	21	56 64
69	6051	6187	6325	6464	6605	6746	6889	7034	7179	7326	20°	63 72
70°	2.7475	2.7625	2.7776	2.7929	2.8083	2.8239	2.8397	2.8556	2.8716	2.8878	19	90
71	2.9042	2.9208	2.9375	2.9544	2.9714	2.9887	3.0061	3.0237	3.0415	3.0595	18	9
72	3.0777	3.0961	3.1146	3.1334	3.1524	3.1716	3.1910	3.2106	3.2305	3.2506	17	18
73	2709	2914	3122	3332	3544	3759	3977	4197	4420	4646	16	27
74	4874	5105	5339	5576	5816	6059	6305	6554	6806	7062	15	36
75	3.7321	3.7583	3.7848	3.8118	3.8391	3.8667	3.8947	3.9232	3.9520	3.9812	14	45
76	4.0108	4.0408	4.0713	4.1022	4.1335	4.1653	4.1976	4.2303	4.2635	4.2972	13	54
77	4.3315	4.3662	4.4015	4.4374	4.4737	4.5107	4.5483	4.5864	4.6252	4.6646	12	63
78	4.7046	4.7453	4.7867	4.8288	4.8716	4.9152	4.9594	5.0045	5.0504	5.0970	11	72
79	5.1446	5.1929	5.2422	5.2924	5.3435	5.3955	5.4486	5.5026	5.5578	5.6140	10°	81
80°	5.6713	5.7297	5.7894	5.8502	5.9124	5.9758	6.0405	6.1066	6.1742	6.2432	9	
81	6.3138	6.3859	6.4596	6.5350	6.6122	6.6912	6.7720	6.8548	6.9395	7.0264	8	
82	7.1154	7.2066	7.3002	7.3962	7.4947	7.5958	7.6996	7.8062	7.9158	8.0285	7	
83	8.1443	8.2636	8.3863	8.5126	8.6427	8.7769	8.9152	9.0579	9.2052	9.3572	6	
84	9.5144	9.677	9.845	10.02	10.20	10.39	10.58	10.78	10.99	11.20	5	
85	11.43	11.66	11.91	12.16	12.43	12.71	13.00	13.30	13.62	13.95	4	
86	14.30	14.67	15.06	15.46	15.89	16.35	16.83	17.34	17.89	18.46	3	
87	19.08	19.74	20.45	21.20	22.02	22.90	23.86	24.90	26.03	27.27	2	
88	28.64	30.14	31.82	33.69	35.80	38.19	40.92	44.07	47.74	52.08	1	
89	57.29	63.66	71.62	81.85	95.49	114.6	143.2	191.0	286.5	573.0	0°	
	1°.0	°.9	°.8	°.7	°.6	°.5	°.4	°.3	°.2	°.1	Deg.	Interpola. for h'dths

4 PL. NAT. COT.

LOGARITHMS OF SINES AND COSINES

TO

FOUR PLACES.

Note. For log. cos. use right-hand column of degrees and lower line of tenths.

Deg.	°.0	°.1	°.2	°.3	°.4	°.5	°.6	°.7	°.8	°.9		Interpola. for h'dths	
0°	−∞	$\bar{3}$.2419	$\bar{3}$.5429	$\bar{3}$.7190	$\bar{3}$.8439	$\bar{3}$.9408	$\bar{2}$.0200	$\bar{2}$.0870	$\bar{2}$.1450	$\bar{2}$.1961	89	35	25
1	$\bar{2}$.2419	$\bar{2}$.2832	$\bar{2}$.3210	$\bar{2}$.3558	$\bar{2}$.3880	$\bar{2}$.4179	$\bar{2}$.4459	$\bar{2}$.4723	$\bar{2}$.4971	$\bar{2}$.5206	88	4	3
2	$\bar{2}$.5428	$\bar{2}$.5640	$\bar{2}$.5842	$\bar{2}$.6035	$\bar{2}$.6220	$\bar{2}$.6397	$\bar{2}$.6567	$\bar{2}$.6731	$\bar{2}$.6889	$\bar{2}$.7041	87	7	5
3	$\bar{2}$.7188	$\bar{2}$.7330	$\bar{2}$.7468	$\bar{2}$.7602	$\bar{2}$.7731	$\bar{2}$.7857	$\bar{2}$.7979	$\bar{2}$.8098	$\bar{2}$.8213	$\bar{2}$.8326	86	11	8
4	$\bar{2}$.8436	$\bar{2}$.8543	$\bar{2}$.8647	$\bar{2}$.8749	$\bar{2}$.8849	$\bar{2}$.8946	$\bar{2}$.9042	$\bar{2}$.9135	$\bar{2}$.9226	$\bar{2}$.9315	85	14	10
5	$\bar{2}$.9403	$\bar{2}$.9489	$\bar{2}$.9573	$\bar{2}$.9655	$\bar{2}$.9736	$\bar{2}$.9816	$\bar{2}$.9894	$\bar{2}$.9970	$\bar{1}$.0046	$\bar{1}$.0120	84	18	13
6	$\bar{1}$.0192	$\bar{1}$.0264	$\bar{1}$.0334	$\bar{1}$.0403	$\bar{1}$.0472	$\bar{1}$.0539	$\bar{1}$.0605	$\bar{1}$.0670	$\bar{1}$.0734	$\bar{1}$.0797	83	21	15
7	$\bar{1}$.0859	$\bar{1}$.0920	$\bar{1}$.0981	$\bar{1}$.1040	$\bar{1}$.1099	$\bar{1}$.1157	$\bar{1}$.1214	$\bar{1}$.1271	$\bar{1}$.1326	$\bar{1}$.1381	82	25	18
8	$\bar{1}$.1436	$\bar{1}$.1489	$\bar{1}$.1542	$\bar{1}$.1594	$\bar{1}$.1646	$\bar{1}$.1697	$\bar{1}$.1747	$\bar{1}$.1797	$\bar{1}$.1847	$\bar{1}$.1895	81	28	20
9	$\bar{1}$.1943	$\bar{1}$.1991	$\bar{1}$.2038	$\bar{1}$.2085	$\bar{1}$.2131	$\bar{1}$.2176	$\bar{1}$.2221	$\bar{1}$.2266	$\bar{1}$.2310	$\bar{1}$.2353	80°	32	23
10°	$\bar{1}$.2397	$\bar{1}$.2439	$\bar{1}$.2482	$\bar{1}$.2524	$\bar{1}$.2565	$\bar{1}$.2606	$\bar{1}$.2647	$\bar{1}$.2687	$\bar{1}$.2727	$\bar{1}$.2767	79	21	18
11	2806	2845	2883	2921	2959	2997	3034	3070	3107	3143	78	2	2
12	3179	3214	3250	3284	3319	3353	3387	3421	3455	3488	77	4	4
13	3521	3554	3586	3618	3650	3682	3713	3745	3775	3806	76	6	5
14	3837	3867	3897	3927	3957	3986	4015	4044	4073	4102	75	8	7
15	$\bar{1}$.4130	$\bar{1}$.4158	$\bar{1}$.4186	$\bar{1}$.4214	$\bar{1}$.4242	$\bar{1}$.4269	$\bar{1}$.4296	$\bar{1}$.4323	$\bar{1}$.4350	$\bar{1}$.4377	74	11	9
16	4403	4430	4456	4482	4508	4533	4559	4584	4609	4634	73	13	11
17	4659	4684	4709	4733	4757	4781	4805	4829	4853	4876	72	15	13
18	4900	4923	4946	4969	4992	5015	5037	5060	5082	5104	71	17	14
19	5126	5148	5170	5192	5213	5235	5256	5278	5299	5320	70°	19	16
20°	$\bar{1}$.5341	$\bar{1}$.5361	$\bar{1}$.5382	$\bar{1}$.5402	$\bar{1}$.5423	$\bar{1}$.5443	$\bar{1}$.5463	$\bar{1}$.5484	$\bar{1}$.5504	$\bar{1}$.5523	69	15	13
21	5543	5563	5583	5602	5621	5641	5660	5679	5698	5717	68	2	1
22	5736	5754	5773	5792	5810	5828	5847	5865	5883	5901	67	3	3
23	5919	5937	5954	5972	5990	6007	6024	6042	6059	6076	66	5	4
24	6093	6110	6127	6144	6161	6177	6194	6210	6227	6243	65	6	5
25	$\bar{1}$.6259	$\bar{1}$.6276	$\bar{1}$.6292	$\bar{1}$.6308	$\bar{1}$.6324	$\bar{1}$.6340	$\bar{1}$.6356	$\bar{1}$.6371	$\bar{1}$.6387	$\bar{1}$.6403	64	8	7
26	6418	6434	6449	6465	6480	6495	6510	6526	6541	6556	63	9	8
27	6570	6585	6600	6615	6629	6644	6659	6673	6687	6702	62	11	9
28	6716	6730	6744	6759	6773	6787	6801	6814	6828	6842	61	12	10
29	6856	6869	6883	6896	6910	6923	6937	6950	6963	6977	60°	14	12
30°	$\bar{1}$.6990	$\bar{1}$.7003	$\bar{1}$.7016	$\bar{1}$.7029	$\bar{1}$.7042	$\bar{1}$.7055	$\bar{1}$.7068	$\bar{1}$.7080	$\bar{1}$.7093	$\bar{1}$.7106	59	11	9
31	7118	7131	7144	7156	7168	7181	7193	7205	7218	7230	58	1	1
32	7242	7254	7266	7278	7290	7302	7314	7326	7338	7349	57	2	2
33	7361	7373	7384	7396	7407	7419	7430	7442	7453	7464	56	3	3
34	7476	7487	7498	7509	7520	7531	7542	7553	7564	7575	55	4	4
35	$\bar{1}$.7586	$\bar{1}$.7597	$\bar{1}$.7607	$\bar{1}$.7618	$\bar{1}$.7629	$\bar{1}$.7640	$\bar{1}$.7650	$\bar{1}$.7661	$\bar{1}$.7671	$\bar{1}$.7682	54	6	5
36	7692	7703	7713	7723	7734	7744	7754	7764	7774	7785	53	7	5
37	7795	7805	7815	7825	7835	7844	7854	7864	7874	7884	52	8	6
38	7893	7903	7913	7922	7932	7941	7951	7960	7970	7979	51	9	7
39	7989	7998	8007	8017	8026	8035	8044	8053	8063	8072	50°	10	8
	1°.0	°.9	°.8	°.7	°.6	°.5	°.4	°.3	°.2	°.1	Deg.	Interpola. for h'dths	

LOGARITHMS OF SINES AND COSINES.

4 PL. LOG. SIN.

Deg.	°.0	°.1	°.2	°.3	°.4	°.5	°.6	°.7	°.8	°.9		Interpola. tables.
40°	$\bar{1}$.8081	$\bar{1}$.8090	$\bar{1}$.8099	$\bar{1}$.8108	$\bar{1}$.8117	$\bar{1}$.8125	$\bar{1}$.8134	$\bar{1}$.8143	$\bar{1}$.8152	$\bar{1}$.8161	49	**9 8 7**
41	8169	8178	8187	8195	8204	8213	8221	8230	8238	8247	48	1 1 1
42	8255	8264	8272	8280	8289	8297	8305	8313	8322	8330	47	2 2 1
43	8338	8346	8354	8362	8370	8378	8386	8394	8402	8410	46	3 2 2
44	8418	8426	8433	8441	8449	8457	8464	8472	8480	8487	45	4 3 3
45	$\bar{1}$.8495	$\bar{1}$.8502	$\bar{1}$.8510	$\bar{1}$.8517	$\bar{1}$.8525	$\bar{1}$.8532	$\bar{1}$.8540	$\bar{1}$.8547	$\bar{1}$.8555	$\bar{1}$.8562	44	5 4 4
46	8569	8577	8584	8591	8598	8606	8613	8620	8627	8634	43	5 5 4
47	8641	8648	8655	8662	8669	8676	8683	8690	8697	8704	42	6 6 5
48	8711	8718	8724	8731	8738	8745	8751	8758	8765	8771	41	7 6 6
49	8778	8784	8791	8797	8804	8810	8817	8823	8830	8836	40°	8 7 6
50°	$\bar{1}$.8843	$\bar{1}$.8849	$\bar{1}$.8855	$\bar{1}$.8862	$\bar{1}$.8868	$\bar{1}$.8874	$\bar{1}$.8880	$\bar{1}$.8887	$\bar{1}$.8893	$\bar{1}$.8899	39	**6 5 4**
51	8905	8911	8917	8923	8929	8935	8941	8947	8953	8959	38	1 1 0
52	8965	8971	8977	8983	8989	8995	9000	9006	9012	9018	37	1 1 1
53	9023	9029	9035	9041	9046	9052	9057	9063	9069	9074	36	2 2 1
54	9080	9085	9091	9096	9101	9107	9112	9118	9123	9128	35	2 2 2
55	$\bar{1}$.9134	$\bar{1}$.9139	$\bar{1}$.9144	$\bar{1}$.9149	$\bar{1}$.9155	$\bar{1}$.9160	$\bar{1}$.9165	$\bar{1}$.9170	$\bar{1}$.9175	$\bar{1}$.9181	34	3 3 2
56	9186	9191	9196	9201	9206	9211	9216	9221	9226	9231	33	4 3 2
57	9236	9241	9246	9251	9255	9260	9265	9270	9275	9279	32	4 4 3
58	9284	9289	9294	9298	9303	9308	9312	9317	9322	9326	31	5 4 3
59	9331	9335	9340	9344	9349	9353	9358	9362	9367	9371	30°	5 5 4
60°	$\bar{1}$.9375	$\bar{1}$.9380	$\bar{1}$.9384	$\bar{1}$.9388	$\bar{1}$.9393	$\bar{1}$.9397	$\bar{1}$.9401	$\bar{1}$.9406	$\bar{1}$.9410	$\bar{1}$.9414	29	**4 3 2**
61	9418	9422	9427	9431	9435	9439	9443	9447	9451	9455	28	0 0 0
62	9459	9463	9467	9471	9475	9479	9483	9487	9491	9495	27	1 1 0
63	9499	9503	9506	9510	9514	9518	9522	9525	9529	9533	26	1 1 1
64	9537	9540	9544	9548	9551	9555	9558	9562	9566	9569	25	2 1 1
65	$\bar{1}$.9573	$\bar{1}$.9576	$\bar{1}$.9580	$\bar{1}$.9583	$\bar{1}$.9587	$\bar{1}$.9590	$\bar{1}$.9594	$\bar{1}$.9597	$\bar{1}$.9601	$\bar{1}$.9604	24	2 2 1
66	9607	9611	9614	9617	9621	9624	9627	9631	9634	9637	23	2 2 1
67	9640	9643	9647	9650	9653	9656	9659	9662	9666	9669	22	3 2 1
68	9672	9675	9678	9681	9684	9687	9690	9693	9696	9699	21	3 2 2
69	9702	9704	9707	9710	9713	9716	9719	9722	9724	9727	20°	4 3 2
70°	$\bar{1}$.9730	$\bar{1}$.9733	$\bar{1}$.9735	$\bar{1}$.9738	$\bar{1}$.9741	$\bar{1}$.9743	$\bar{1}$.9746	$\bar{1}$.9749	$\bar{1}$.9751	$\bar{1}$.9754	19	**3 2 1**
71	9757	9759	9762	9764	9767	9770	9772	9775	9777	9780	18	0 0 0
72	9782	9785	9787	9789	9792	9794	9797	9799	9801	9804	17	1 0 0
73	9806	9808	9811	9813	9815	9817	9820	9822	9824	9826	16	1 1 0
74	9828	9831	9833	9835	9837	9839	9841	9843	9845	9847	15	1 1 0
75	$\bar{1}$.9849	$\bar{1}$.9851	$\bar{1}$.9853	$\bar{1}$.9855	$\bar{1}$.9857	$\bar{1}$.9859	$\bar{1}$.9861	$\bar{1}$.9863	$\bar{1}$.9865	$\bar{1}$.9867	14	2 1 1
76	9869	9871	9873	9875	9876	9878	9880	9882	9884	9885	13	2 1 1
77	9887	9889	9891	9892	9894	9896	9897	9899	9901	9902	12	2 1 1
78	9904	9906	9907	9909	9910	9912	9913	9915	9916	9918	11	2 2 1
79	9919	9921	9922	9924	9925	9927	9928	9929	9931	9932	10°	3 2 1
80°	$\bar{1}$.9934	$\bar{1}$.9935	$\bar{1}$.9936	$\bar{1}$.9937	$\bar{1}$.9939	$\bar{1}$.9940	$\bar{1}$.9941	$\bar{1}$.9943	$\bar{1}$.9944	$\bar{1}$.9945	9	**2 1**
81	9946	9947	9949	9950	9951	9952	9953	9954	9955	9956	8	0 0
82	9958	9959	9960	9961	9962	9963	9964	9965	9966	9967	7	0 0
83	9968	9968	9969	9970	9971	9972	9973	9974	9975	9975	6	1 0
84	9976	9977	9978	9978	9979	9980	9981	9981	9982	9983	5	1 0
85	$\bar{1}$.9983	$\bar{1}$.9984	$\bar{1}$.9985	$\bar{1}$.9985	$\bar{1}$.9986	$\bar{1}$.9987	$\bar{1}$.9987	$\bar{1}$.9988	$\bar{1}$.9988	$\bar{1}$.9989	4	1 1
86	9989	9990	9990	9991	9991	9992	9992	9993	9993	9994	3	1 1
87	9994	9994	9995	9995	9996	9996	9996	9996	9997	9997	2	1 1
88	9997	9998	9998	9998	9998	9999	9999	9999	9999	9999	1	2 1
89	9999	9999	zero	zero	zero	zero	zero	zero	zero	zero	0°	2 1
	1°.0	°.9	°.8	°.7	°.6	°.5	°.4	°.3	°.2	°.1	Deg.	Interpola. for h'dths

4 PL. LOG. COS.

LOGARITHMS OF TANGENTS AND COTANGENTS
TO
FOUR PLACES.

Note. For log. cot. use right-hand column of degrees and lower line of tenth

Deg.	°.0	°.1	°.2	°.3	°.4	°.5	°.6	°.7	°.8	°.9		Interpola. for h'dths	
0°	$-\infty$	$\bar{3}.2419$	$\bar{3}.5429$	$\bar{3}.7190$	$\bar{3}.8439$	$\bar{3}.9409$	$\bar{2}.0200$	$\bar{2}.0870$	$\bar{2}.1450$	$\bar{2}.1962$	89	**35**	**27**
1	$\bar{2}.2419$	$\bar{2}.2833$	$\bar{2}.3211$	$\bar{2}.3559$	$\bar{2}.3881$	$\bar{2}.4181$	$\bar{2}.4461$	$\bar{2}.4725$	$\bar{2}.4973$	$\bar{2}.5208$	88	4	3
2	$\bar{2}.5431$	$\bar{2}.5643$	$\bar{2}.5845$	$\bar{2}.6038$	$\bar{2}.6223$	$\bar{2}.6401$	$\bar{2}.6571$	$\bar{2}.6736$	$\bar{2}.6894$	$\bar{2}.7046$	87	7	5
3	$\bar{2}.7194$	$\bar{2}.7337$	$\bar{2}.7475$	$\bar{2}.7609$	$\bar{2}.7739$	$\bar{2}.7865$	$\bar{2}.7988$	$\bar{2}.8107$	$\bar{2}.8223$	$\bar{2}.8336$	86	11	8
4	$\bar{2}.8446$	$\bar{2}.8554$	$\bar{2}.8659$	$\bar{2}.8762$	$\bar{2}.8862$	$\bar{2}.8960$	$\bar{2}.9056$	$\bar{2}.9150$	$\bar{2}.9241$	$\bar{2}.9331$	85	14	11
5	$\bar{2}.9420$	$\bar{2}.9506$	$\bar{2}.9591$	$\bar{2}.9674$	$\bar{2}.9756$	$\bar{2}.9836$	$\bar{2}.9915$	$\bar{2}.9992$	$\bar{1}.0068$	$\bar{1}.0143$	84	18	14
6	$\bar{1}.0216$	$\bar{1}.0289$	$\bar{1}.0360$	$\bar{1}.0430$	$\bar{1}.0499$	$\bar{1}.0567$	$\bar{1}.0633$	$\bar{1}.0699$	$\bar{1}.0764$	$\bar{1}.0828$	83	21	16
7	$\bar{1}.0891$	$\bar{1}.0954$	$\bar{1}.1015$	$\bar{1}.1076$	$\bar{1}.1135$	$\bar{1}.1194$	$\bar{1}.1252$	$\bar{1}.1310$	$\bar{1}.1367$	$\bar{1}.1423$	82	25	19
8	$\bar{1}.1478$	$\bar{1}.1533$	$\bar{1}.1587$	$\bar{1}.1640$	$\bar{1}.1693$	$\bar{1}.1745$	$\bar{1}.1797$	$\bar{1}.1848$	$\bar{1}.1898$	$\bar{1}.1948$	81	28	22
9	$\bar{1}.1997$	$\bar{1}.2046$	$\bar{1}.2094$	$\bar{1}.2142$	$\bar{1}.2189$	$\bar{1}.2236$	$\bar{1}.2282$	$\bar{1}.2328$	$\bar{1}.2374$	$\bar{1}.2419$	80°	32	24
10°	$\bar{1}.2463$	$\bar{1}.2507$	$\bar{1}.2551$	$\bar{1}.2594$	$\bar{1}.2637$	$\bar{1}.2680$	$\bar{1}.2722$	$\bar{1}.2764$	$\bar{1}.2805$	$\bar{1}.2846$	79	**25**	**23**
11	2887	2927	2967	3006	3046	3085	3123	3162	3200	3237	78	3	2
12	3275	3312	3349	3385	3422	3458	3493	3529	3564	3599	77	5	5
13	3634	3668	3702	3736	3770	3804	3837	3870	3903	3935	76	8	7
14	3968	4000	4032	4064	4095	4127	4158	4189	4220	4250	75	10	9
15	$\bar{1}.4281$	$\bar{1}.4311$	$\bar{1}.4341$	$\bar{1}.4371$	$\bar{1}.4400$	$\bar{1}.4430$	$\bar{1}.4459$	$\bar{1}.4488$	$\bar{1}.4517$	$\bar{1}.4546$	74	13	12
16	4575	4603	4632	4660	4688	4716	4744	4771	4799	4826	73	15	14
17	4853	4880	4907	4934	4961	4987	5014	5040	5066	5092	72	18	16
18	5118	5143	5169	5195	5220	5245	5270	5295	5320	5345	71	20	18
19	5370	5394	5419	5443	5467	5491	5516	5539	5563	5587	70°	23	21
20°	$\bar{1}.5611$	$\bar{1}.5634$	$\bar{1}.5658$	$\bar{1}.5681$	$\bar{1}.5704$	$\bar{1}.5727$	$\bar{1}.5750$	$\bar{1}.5773$	$\bar{1}.5796$	$\bar{1}.5819$	69	**21**	**19**
21	5842	5864	5887	5909	5932	5954	5976	5998	6020	6042	68	2	2
22	6064	6086	6108	6129	6151	6172	6194	6215	6236	6257	67	4	4
23	6279	6300	6321	6341	6362	6383	6404	6424	6445	6465	66	6	6
24	6486	6506	6527	6547	6567	6587	6607	6627	6647	6667	65	8	8
25	$\bar{1}.6687$	$\bar{1}.6706$	$\bar{1}.6726$	$\bar{1}.6746$	$\bar{1}.6765$	$\bar{1}.6785$	$\bar{1}.6804$	$\bar{1}.6824$	$\bar{1}.6843$	$\bar{1}.6863$	64	11	10
26	6882	6901	6920	6939	6958	6977	6996	7015	7034	7053	63	13	11
27	7072	7090	7109	7128	7146	7165	7183	7202	7220	7238	62	15	13
28	7257	7275	7293	7311	7330	7348	7366	7384	7402	7420	61	17	15
29	7438	7455	7473	7491	7509	7526	7544	7562	7579	7597	60°	19	17
30°	$\bar{1}.7614$	$\bar{1}.7632$	$\bar{1}.7649$	$\bar{1}.7667$	$\bar{1}.7684$	$\bar{1}.7701$	$\bar{1}.7719$	$\bar{1}.7736$	$\bar{1}.7753$	$\bar{1}.7771$	59	**17**	**15**
31	7788	7805	7822	7839	7856	7873	7890	7907	7924	7941	58	2	2
32	7958	7975	7992	8008	8025	8042	8059	8075	8092	8109	57	3	3
33	8125	8142	8158	8175	8191	8208	8224	8241	8257	8274	56	5	5
34	8290	8306	8323	8339	8355	8371	8388	8404	8420	8436	55	7	6
35	$\bar{1}.8452$	$\bar{1}.8468$	$\bar{1}.8484$	$\bar{1}.8501$	$\bar{1}.8517$	$\bar{1}.8533$	$\bar{1}.8549$	$\bar{1}.8565$	$\bar{1}.8581$	$\bar{1}.8597$	54	9	8
36	8613	8629	8644	8660	8676	8692	8708	8724	8740	8755	53	10	9
37	8771	8787	8803	8818	8834	8850	8865	8881	8897	8912	52	12	11
38	8928	8944	8959	8975	8990	9006	9022	9037	9053	9068	51	14	12
39	9084	9099	9115	9130	9146	9161	9176	9192	9207	9223	50°	15	14
	1°.0	°.9	°.8	°.7	°.6	°.5	°.4	°.3	°.2	°.1	Deg.	Interpola. for h'dths	

Deg.	°.0	°.1	°.2	°.3	°.4	°.5	°.6	°.7	°.8	°.9		Interpola. for h'dths	
40°	$\bar{1}.9238$	$\bar{1}.9254$	$\bar{1}.9269$	$\bar{1}.9284$	$\bar{1}.9300$	$\bar{1}.9315$	$\bar{1}.9330$	$\bar{1}.9346$	$\bar{1}.9361$	$\bar{1}.9376$	49	**15**	**16**
41	$\bar{1}.9392$	$\bar{1}.9407$	$\bar{1}.9422$	$\bar{1}.9438$	$\bar{1}.9453$	$\bar{1}.9468$	$\bar{1}.9483$	$\bar{1}.9499$	$\bar{1}.9514$	$\bar{1}.9529$	48	2	2
42	$\bar{1}.9544$	$\bar{1}.9560$	$\bar{1}.9575$	$\bar{1}.9590$	$\bar{1}.9605$	$\bar{1}.9621$	$\bar{1}.9636$	$\bar{1}.9651$	$\bar{1}.9666$	$\bar{1}.9681$	47	3	3
43	$\bar{1}.9697$	$\bar{1}.9712$	$\bar{1}.9727$	$\bar{1}.9742$	$\bar{1}.9757$	$\bar{1}.9772$	$\bar{1}.9788$	$\bar{1}.9803$	$\bar{1}.9818$	$\bar{1}.9833$	46	5	5
44	$\bar{1}.9848$	$\bar{1}.9864$	$\bar{1}.9879$	$\bar{1}.9894$	$\bar{1}.9909$	$\bar{1}.9924$	$\bar{1}.9939$	$\bar{1}.9955$	$\bar{1}.9970$	$\bar{1}.9985$	45	6	6
45	0.0000	0.0015	0.0030	0.0045	0.0061	0.0076	0.0091	0.0106	0.0121	0.0136	44	8	8
46	0152	0167	0182	0197	0212	0228	0243	0258	0273	0288	43	9	10
47	0303	0319	0334	0349	0364	0379	0395	0410	0425	0440	42	11	11
48	0456	0471	0486	0501	0517	0532	0547	0562	0578	0593	41	12	13
49	0608	0624	0639	0654	0670	0685	0700	0716	0731	0746	40°	14	14
50°	0.0762	0.0777	0.0793	0.0808	0.0824	0.0839	0.0854	0.0870	0.0885	0.0901	39	**17**	**18**
51	0916	0932	0947	0963	0978	0994	1010	1025	1041	1056	38	2	2
52	1072	1088	1103	1119	1135	1150	1166	1182	1197	1213	37	3	4
53	1229	1245	1260	1276	1292	1308	1324	1340	1356	1371	36	5	5
54	1387	1403	1419	1435	1451	1467	1483	1499	1516	1532	35	7	7
55	0.1548	0.1564	0.1580	0.1596	0.1612	0.1629	0.1645	0.1661	0.1677	0.1694	34	9	9
56	1710	1726	1743	1759	1776	1792	1809	1825	1842	1858	33	10	11
57	1875	1891	1908	1925	1941	1958	1975	1992	2008	2025	32	12	13
58	2042	2059	2076	2093	2110	2127	2144	2161	2178	2195	31	14	14
59	2212	2229	2247	2264	2281	2299	2316	2333	2351	2368	30°	15	16
60°	0.2386	0.2403	0.2421	0.2438	0.2456	0.2474	0.2491	0.2509	0.2527	0.2545	29	**21**	**23**
61	2562	2580	2598	2616	2634	2652	2670	2689	2707	2725	28	2	2
62	2743	2762	2780	2798	2817	2835	2854	2872	2891	2910	27	4	5
63	2928	2947	2966	2985	3004	3023	3042	3061	3080	3099	26	6	7
64	3118	3137	3157	3176	3196	3215	3235	3254	3274	3294	25	8	9
65	0.3313	0.3333	0.3353	0.3373	0.3393	0.3413	0.3433	0.3453	0.3473	0.3494	24	11	12
66	3514	3535	3555	3576	3596	3617	3638	3659	3679	3700	23	13	14
67	3721	3743	3764	3785	3806	3828	3849	3871	3892	3914	22	15	16
68	3936	3958	3980	4002	4024	4046	4068	4091	4113	4136	21	17	18
69	4158	4181	4204	4227	4250	4273	4296	4319	4342	4366	20°	19	21
70°	0.4389	0.4413	0.4437	0.4461	0.4484	0.4509	0.4533	0.4557	0.4581	0.4606	19	**25**	**27**
71	4630	4655	4680	4705	4730	4755	4780	4805	4831	4857	18	3	3
72	4882	4908	4934	4960	4986	5013	5039	5066	5093	5120	17	5	5
73	5147	5174	5201	5229	5256	5284	5312	5340	5368	5397	16	8	8
74	5425	5454	5483	5512	5541	5570	5600	5629	5659	5689	15	10	11
75	0.5719	0.5750	0.5780	0.5811	0.5842	0.5873	0.5905	0.5936	0.5968	0.6000	14	13	14
76	6032	6065	6097	6130	6163	6196	6230	6264	6298	6332	13	15	16
77	6366	6401	6436	6471	6507	6542	6578	6615	6651	6688	12	18	19
78	6725	6763	6800	6838	6877	6915	6954	6994	7033	7073	11	20	22
79	7113	7154	7195	7236	7278	7320	7363	7406	7449	7493	10°	23	24
80°	0.7537	0.7581	0.7626	0.7672	0.7718	0.7764	0.7811	0.7858	0.7906	0.7954	9	**35**	**45**
81	0.8003	0.8052	0.8102	0.8152	0.8203	0.8255	0.8307	0.8360	0.8413	0.8467	8	4	5
82	0.8522	0.8577	0.8633	0.8690	0.8748	0.8806	0.8865	0.8924	0.8985	0.9046	7	7	9
83	0.9109	0.9172	0.9236	0.9301	0.9367	0.9433	0.9501	0.9570	0.9640	0.9711	6	11	14
84	0.9784	0.9857	0.9932	1.0008	1.0085	1.0164	1.0244	1.0326	1.0409	1.0494	5	14	18
85	1.0580	1.0669	1.0759	1.0850	1.0944	1.1040	1.1138	1.1238	1.1341	1.1446	4	18	23
86	1.1554	1.1664	1.1777	1.1893	1.2012	1.2135	1.2261	1.2391	1.2525	1.2663	3	21	27
87	1.2806	1.2954	1.3106	1.3264	1.3429	1.3599	1.3777	1.3962	1.4155	1.4357	2	25	32
88	1.4569	1.4792	1.5027	1.5275	1.5539	1.5819	1.6119	1.6441	1.6789	1.7167	1	28	36
89	1.7581	1.8038	1.8550	1.9130	1.9800	2.0591	2.1561	2.2810	2.4571	2.7581	0°	32	41
	1°.0	°.9	°.8	°.7	°.6	°.5	°.4	°.3	°.2	°.1	Deg.	Interpola. for h'dths	

LOGARITHMS OF TRIGONOMETRIC FUNCTIONS

TO

FIVE PLACES.

Note. The table gives the log of the natural value of the function, and hence the characteristic is negative when that value is fractional. The common practice of adding 10. to avoid the negative characteristic is not recommended.

0°	log cos		0°	log cos	
0′–16′	0.00 000	44′–60′	44′–49′	$\bar{1}$.99 996	11′–16′
17′–28′	$\bar{1}$.99 999	32′–43′	50′–54′	$\bar{1}$.99 995	6′–10′
29′–36′	$\bar{1}$.99 998	24′–31′	55′–59′	$\bar{1}$.99 994	1′–5′
37′–43′	$\bar{1}$.99 997	17′–23′	60′	$\bar{1}$.99 993	0′
	log sin	89°		log sin	89°

0° ′	log sin	log tan	log cot		0° ′	log sin	log tan	log cot	
0′	$\bar{\infty}$	$\bar{\infty}$	∞	60′	30′	$\bar{3}$.94 084	$\bar{3}$.94 086	2.05 914	30′
1	$\bar{4}$.46 373	$\bar{4}$.46 373	3.53 627	59	31	95 508	95 510	04 490	29
2	$\bar{4}$.76 476	$\bar{4}$.76 476	3.23 524	58	32	96 887	96 889	03 111	28
3	$\bar{4}$.94 085	$\bar{4}$.94 085	3.05 915	57	33	98 223	98 225	01 775	27
4	$\bar{3}$.06 579	$\bar{3}$.06 579	2.93 421	56	34	$\bar{3}$.99 520	$\bar{3}$.99 522	2.00 478	26
5	$\bar{3}$.16 270	$\bar{3}$.16 270	2.83 730	55	35	$\bar{2}$.00 779	$\bar{2}$.00 781	1.99 219	25
6	24 188	24 188	75 812	54	36	02 002	02 004	97 996	24
7	30 882	30 822	69 118	53	37	03 192	03 194	96 806	23
8	36 682	36 682	63 318	52	38	04 350	04 353	95 647	22
9	41 797	41 797	58 203	51	39	05 478	05 481	94 519	21
10	$\bar{3}$.46 373	$\bar{3}$.46 373	2.53 627	50	40	$\bar{2}$.06 578	$\bar{2}$.06 581	1.93 419	20
11	50 512	50 512	49 488	49	41	07 650	07 653	92 347	19
12	54 291	54 291	45 709	48	42	08 696	08 700	91 300	18
13	57 767	57 767	42 233	47	43	09 718	09 722	90 278	17
14	60 985	60 986	39 014	46	44	10 717	10 720	89 280	16
15	$\bar{3}$.63 982	$\bar{3}$.63 982	2.36 018	45	45	$\bar{2}$.11 693	2.11 696	1.88 304	15
16	66 784	66 785	33 215	44	46	12 647	12 651	87 349	14
17	69 417	69 418	30 582	43	47	13 581	13 585	86 415	13
18	71 900	71 900	28 100	42	48	14 495	14 500	85 500	12
19	74 248	74 248	25 752	41	49	15 391	15 395	84 605	11
20	$\bar{3}$.76 475	$\bar{3}$.76 476	2.23 524	40	50	$\bar{2}$.16 268	$\bar{2}$.16 273	1.83 727	10
21	78 594	78 595	21 405	39	51	17 128	17 133	82 867	9
22	80 615	80 615	19 385	38	52	17 971	17 976	82 024	8
23	82 545	82 546	17 454	37	53	18 798	18 804	81 196	7
24	84 393	84 394	15 606	36	54	19 610	19 616	80 384	6
25	$\bar{3}$.86 166	$\bar{3}$.86 167	2.13 833	35	55	$\bar{2}$.20 407	$\bar{2}$.20 413	1.79 587	5
26	87 870	87 871	12 129	34	56	21 189	21 195	78 805	4
27	89 509	89 510	10 490	33	57	21 958	21 964	78 036	3
28	91 088	91 089	08 911	32	58	22 713	22 720	77 280	2
29	92 612	92 613	07 387	31	59	23 456	23 462	76 538	1
30′	$\bar{3}$.94 084	$\bar{3}$.94 086	2.05 914	30′	60′	$\bar{2}$.24 186	$\bar{2}$.24 192	1.75 808	0′
	log cos	log cot	log tan	′ 89°		log cos	log cot	log tan	′ 89°

′	log sin	log tan	log cot	log cos	log sin	log tan	log cot	log cos	
0′	$\bar{2}$.24 186	$\bar{2}$.24 192	1.75 808	$\bar{1}$.99 993	$\bar{2}$.54 282	$\bar{2}$.54 308	1.45 692	$\bar{1}$.99 974	60′
1	24 903	24 910	75 090	99 993	54 642	54 669	45 331	99 973	59
2	25 609	25 616	74 384	99 993	54 999	55 027	44 973	99 973	58
3	26 304	26 312	73 688	99 993	55 354	55 382	44 618	99 972	57
4	26 988	26 996	73 004	99 992	55 705	55 734	44 266	99 972	56
5	$\bar{2}$.27 661	$\bar{2}$.27 669	1.72 331	$\bar{1}$.99 992	$\bar{2}$.56 054	$\bar{2}$.56 083	1.43 917	$\bar{1}$.99 971	55
6	28 324	28 332	71 668	99 992	56 400	56 429	43 571	99 971	54
7	28 977	28 986	71 014	99 992	56 743	56 773	43 227	99 970	53
8	29 621	29 629	70 371	99 992	57 084	57 114	42 886	99 970	52
9	30 255	30 263	69 737	99 991	57 421	57 452	42 548	99 969	51
10	$\bar{2}$.30 879	$\bar{2}$.30 888	1.69 112	$\bar{1}$.99 991	$\bar{2}$.57 757	$\bar{2}$.57 788	1.42 212	$\bar{1}$.99 969	50
11	31 495	31 505	68 495	99 991	58 089	58 121	41 879	99 968	49
12	32 103	32 112	67 888	99 990	58 419	58 451	41 549	99 968	48
13	32 702	32 711	67 289	99 990	58 747	58 779	41 221	99 967	47
14	33 292	33 302	66 698	99 990	59 072	59 105	40 895	99 967	46
15	$\bar{2}$.33 875	$\bar{2}$.33 886	1.66 114	$\bar{1}$.99 990	$\bar{2}$.59 395	$\bar{2}$.59 428	1.40 572	$\bar{1}$.99 967	45
16	34 450	34 461	65 539	99 989	59 715	59 749	40 251	99 966	44
17	35 018	35 029	64 971	99 989	60 033	60 068	39 932	99 966	43
18	35 578	35 590	64 410	99 989	60 349	60 384	39 616	99 965	42
19	36 131	36 143	63 857	99 989	60 662	60 698	39 302	99 964	41
20	$\bar{2}$.36 678	$\bar{2}$.36 689	1.63 311	$\bar{1}$.99 988	$\bar{2}$.60 973	$\bar{2}$.61 009	1.38 991	$\bar{1}$.99 964	40
21	37 217	37 229	62 771	99 988	61 282	61 319	38 681	99 963	39
22	37 750	37 762	62 238	99 988	61 589	61 626	38 374	99 963	38
23	38 276	38 289	61 711	99 987	61 894	61 931	38 069	99 962	37
24	38 796	38 809	61 191	99 987	62 196	62 234	37 766	99 962	36
25	$\bar{2}$.39 310	$\bar{2}$.39 323	1.60 677	$\bar{1}$.99 987	$\bar{2}$.62 497	$\bar{2}$.62 535	1.37 465	$\bar{1}$.99 961	35
26	39 818	39 832	60 168	99 986	62 795	62 834	37 166	99 961	34
27	40 320	40 334	59 666	99 986	63 091	63 131	36 869	99 960	33
28	40 816	40 830	59 170	99 986	63 385	63 426	36 574	99 960	32
29	41 307	41 321	58 679	99 985	63 678	63 718	36 282	99 959	31
30	$\bar{2}$.41 792	$\bar{2}$.41 807	1.58 193	$\bar{1}$.99 985	$\bar{2}$.63 968	$\bar{2}$.64 009	1.35 991	$\bar{1}$.99 959	30
31	42 272	42 287	57 713	99 985	64 256	64 298	35 702	99 958	29
32	42 746	42 762	57 238	99 984	64 543	64 585	35 415	99 958	28
33	43 216	43 232	56 768	99 984	64 827	64 870	35 130	99 957	27
34	43 680	43 696	56 304	99 984	65 110	65 154	34 846	99 956	26
35	$\bar{2}$.44 139	$\bar{2}$.44 156	1.55 844	$\bar{1}$.99 983	$\bar{2}$.65 391	$\bar{2}$.65 435	1.34 565	$\bar{1}$.99 956	25
36	44 594	44 611	55 389	99 983	65 670	65 715	34 285	99 955	24
37	45 044	45 061	54 939	99 983	65 947	65 993	34 007	99 955	23
38	45 489	45 507	54 493	99 982	66 223	66 269	33 731	99 954	22
39	45 930	45 948	54 052	99 982	66 497	66 543	33 457	99 954	21
40	$\bar{2}$.46 366	$\bar{2}$.46 385	1.53 615	$\bar{1}$.99 982	$\bar{2}$.66 769	$\bar{2}$.66 816	1.33 184	$\bar{1}$.99 953	20
41	46 799	46 817	53 183	99 981	67 039	67 087	32 913	99 952	19
42	47 226	47 245	52 755	99 981	67 308	67 356	32 644	99 952	18
43	47 650	47 669	52 331	99 981	67 575	67 624	32 376	99 951	17
44	48 069	48 089	51 911	99 980	67 841	67 890	32 110	99 951	16
45	$\bar{2}$.48 485	$\bar{2}$.48 505	1.51 495	$\bar{1}$.99 980	$\bar{2}$.68 104	$\bar{2}$.68 154	1.31 846	$\bar{1}$.99 950	15
46	48 896	48 917	51 083	99 979	68 367	68 417	31 583	99 949	14
47	49 304	49 325	50 675	99 979	68 627	68 678	31 322	99 949	13
48	49 708	49 729	50 271	99 979	68 886	68 938	31 062	99 948	12
49	50 108	50 130	49 870	99 978	69 144	69 196	30 804	99 948	11
50	$\bar{2}$.50 504	$\bar{2}$.50 527	1.49 473	$\bar{1}$.99 978	$\bar{2}$.69 400	$\bar{2}$.69 453	1.30 547	$\bar{1}$.99 947	10
51	50 897	50 920	49 080	99 977	69 654	69 708	30 292	99 946	9
52	51 287	51 310	48 690	99 977	69 907	69 962	30 038	99 946	8
53	51 673	51 696	48 304	99 977	70 159	70 214	29 786	99 945	7
54	52 055	52 079	47 921	99 976	70 409	70 465	29 535	99 944	6
55	$\bar{2}$.52 434	$\bar{2}$.52 459	1.47 541	$\bar{1}$.99 976	$\bar{2}$.70 658	$\bar{2}$.70 714	1.29 286	$\bar{1}$.99 944	5
56	52 810	52 835	47 165	99 975	70 905	70 962	29 038	99 943	4
57	53 183	53 208	46 792	99 975	71 151	71 208	28 792	99 942	3
58	53 552	53 578	46 422	99 974	71 395	71 453	28 547	99 942	2
59	53 919	53 945	46 055	99 974	71 638	71 697	28 303	99 941	1
60′	$\bar{2}$.54 282	$\bar{2}$.54 308	1.45 692	$\bar{1}$.99 974	$\bar{2}$.71 880	$\bar{2}$.71 940	1.28 060	$\bar{1}$.99 940	0′
	log cos	log cot	log tan	log sin	log cos	log cot	log tan	log sin	′

	3°				4°				
′	log sin	log tan	log cot	log cos	log sin	log tan	log cot	log cos	
0	2̄.71 880	2̄.71 940	1.28 060	1̄.99 940	2̄.84 358	2̄.84 464	1.15 536	1̄.99 894	60
1	72 120	72 181	27 819	99 940	84 539	84 646	15 354	99 893	59
2	72 359	72 420	27 580	99 939	84 718	84 826	15 174	99 892	58
3	72 597	72 659	27 341	99 938	84 897	85 006	14 994	99 891	57
4	72 834	72 896	27 104	99 938	85 075	85 185	14 815	99 891	56
5	2̄.73 069	2̄.73 132	1.26 868	1̄.99 937	2̄.85 252	2̄.85 363	1.14 637	1̄.99 890	55
6	73 303	73 366	26 634	99 936	85 429	85 540	14 460	99 889	54
7	73 535	73 600	26 400	99 936	85 605	85 717	14 283	99 888	53
8	73 767	73 832	26 168	99 935	85 780	85 893	14 107	99 887	52
9	73 997	74 063	25 937	99 934	85 955	86 069	13 931	99 886	51
10	2̄.74 226	2̄.74 292	1.25 708	1̄.99 934	2̄.86 128	2̄.86 243	1.13 757	1̄.99 885	50
11	74 454	74 521	25 479	99 933	86 301	86 417	13 583	99 884	49
12	74 680	74 748	25 252	99 932	86 474	86 591	13 409	99 883	48
13	74 906	74 974	25 026	99 932	86 645	86 763	13 237	99 882	47
14	75 130	75 199	24 801	99 931	86 816	86 935	13 065	99 881	46
15	2̄.75 353	2̄.75 423	1.24 577	1̄.99 930	2̄.86 987	2̄.87 106	1.12 894	1̄.99 880	45
16	75 575	75 645	24 355	99 929	87 156	87 277	12 723	99 879	44
17	75 795	75 867	24 133	99 929	87 325	87 447	12 553	99 879	43
18	76 015	76 087	23 913	99 928	87 494	87 616	12 384	99 878	42
19	76 234	76 306	23 694	99 927	87 661	87 785	12 215	99 877	41
20	2̄.76 451	2̄.76 525	1.23 475	1̄.99 926	2̄.87 829	2̄.87 953	1.12 047	1̄.99 876	40
21	76 667	76 742	23 258	99 926	87 995	88 120	11 880	99 875	39
22	76 883	76 958	23 042	99 925	88 161	88 287	11 713	99 874	38
23	77 097	77 173	22 827	99 924	88 326	88 453	11 547	99 873	37
24	77 310	77 387	22 613	99 923	88 490	88 618	11 382	99 872	36
25	2̄.77 522	2̄.77 600	1.22 400	1̄.99 923	2̄.88 654	2̄.88 783	1.11 217	1̄.99 871	35
26	77 733	77 811	22 189	99 922	88 817	88 948	11 052	99 870	34
27	77 943	78 022	21 978	99 921	88 980	89 111	10 889	99 869	33
28	78 152	78 232	21 768	99 920	89 142	89 274	10 726	99 868	32
29	78 360	78 441	21 559	99 920	89 304	89 437	10 563	99 867	31
30	2̄.78 568	2̄.78 649	1.21 351	1̄.99 919	2̄.89 464	2̄.89 598	1.10 402	1̄.99 866	30
31	78 774	78 855	21 145	99 918	89 625	89 760	10 240	99 865	29
32	78 979	79 061	20 939	99 917	89 784	89 920	10 080	99 864	28
33	79 183	79 266	20 734	99 917	89 943	90 080	09 920	99 863	27
34	79 386	79 470	20 530	99 916	90 102	90 240	09 760	99 862	26
35	2̄.79 588	2̄.79 673	1.20 327	1̄.99 915	2̄.90 260	2̄.90 399	1.09 601	1̄.99 861	25
36	79 789	79 875	20 125	99 914	90 417	90 557	09 443	99 860	24
37	79 990	80 076	19 924	99 913	90 574	90 715	09 285	99 859	23
38	80 189	80 277	19 723	99 913	90 730	90 872	09 128	99 858	22
39	80 388	80 476	19 524	99 912	90 885	91 029	08 971	99 857	21
40	2̄.80 585	2̄.80 674	1.19 326	1̄.99 911	2̄.91 040	2̄.91 185	1.08 815	1̄.99 856	20
41	80 782	80 872	19 128	99 910	91 195	91 340	08 660	99 855	19
42	80 978	81 068	18 932	99 909	91 349	91 495	08 505	99 854	18
43	81 173	81 264	18 736	99 909	91 502	91 650	08 350	99 853	17
44	81 367	81 459	18 541	99 908	91 655	91 803	08 197	99 852	16
45	2̄.81 560	2̄.81 653	1.18 347	1̄.99 907	2̄.91 807	2̄.91 957	1.08 043	1̄.99 851	15
46	81 752	81 846	18 154	99 906	91 959	92 110	07 890	99 850	14
47	81 944	82 038	17 962	99 905	92 110	92 262	07 738	99 848	13
48	82 134	82 230	17 770	99 904	92 261	92 414	07 586	99 847	12
49	82 324	82 420	17 580	99 904	92 411	92 565	07 435	99 846	11
50	2̄.82 513	2̄.82 610	1.17 390	1̄.99 903	2̄.92 561	2̄.92 716	1.07 284	1̄.99 845	10
51	82 701	82 799	17 201	99 902	92 710	92 866	07 134	99 844	9
52	82 888	82 987	17 013	99 901	92 859	93 016	06 984	99 843	8
53	83 075	83 175	16 825	99 900	93 007	93 165	06 835	99 842	7
54	83 261	83 361	16 639	99 899	93 154	93 313	06 687	99 841	6
55	2̄.83 446	2̄.83 547	1.16 453	1̄.99 898	2̄.93 301	2̄.93 462	1.06 538	1̄.99 840	5
56	83 630	83 732	16 268	99 898	93 448	93 609	06 391	99 839	4
57	83 813	83 916	16 084	99 897	93 594	93 756	06 244	99 838	3
58	83 996	84 100	15 900	99 896	93 740	93 903	06 097	99 837	2
59	84 177	84 282	15 718	99 895	93 885	94 049	05 951	99 836	1
60	2̄.84 358	2̄.84 464	1.15 536	1̄.99 894	2̄.94 030	2̄.94 195	1.05 805	1̄.99 834	0
	log cos	log cot	log tan	log sin	log cos	log cot	log tan	log sin	′
	86°				85°				

′	log sin (5°)	log tan (5°)	log cot (5°)	log cos (5°)	log sin (6°)	log tan (6°)	log cot (6°)	log cos (6°)	
0′	$\bar{2}$.94 030	$\bar{2}$.94 195	1.05 805	$\bar{1}$.99 834	$\bar{1}$.01 923	$\bar{1}$.02 162	0.97 838	$\bar{1}$.99 761	60′
1	94 174	94 340	05 660	99 833	02 043	02 283	97 717	99 760	59
2	94 317	94 485	05 515	99 832	02 163	02 404	97 596	99 759	58
3	94 461	94 630	05 370	99 831	02 283	02 525	97 475	99 757	57
4	94 603	94 773	05 227	99 830	02 402	02 645	97 355	99 756	56
5	$\bar{2}$.94 746	$\bar{2}$.94 917	1.05 083	$\bar{1}$.99 829	$\bar{1}$.02 520	$\bar{1}$.02 766	0.97 234	$\bar{1}$.99 755	55
6	94 887	95 060	04 940	99 828	02 639	02 885	97 115	99 753	54
7	95 029	95 202	04 798	99 827	02 757	03 005	96 995	99 752	53
8	95 170	95 344	04 656	99 825	02 874	03 124	96 876	99 751	52
9	95 310	95 486	04 514	99 824	02 992	03 242	96 758	99 749	51
10	$\bar{2}$.95 450	$\bar{2}$.95 627	1.04 373	$\bar{1}$.99 823	$\bar{1}$.03 109	$\bar{1}$.03 361	0.96 639	$\bar{1}$.99 748	50
11	95 589	95 767	04 233	99 822	03 226	03 479	96 521	99 747	49
12	95 728	95 908	04 092	99 821	03 342	03 597	96 403	99 745	48
13	95 867	96 047	03 953	99 820	03 458	03 714	96 286	99 744	47
14	96 005	96 187	03 813	99 819	03 574	03 832	96 168	99 742	46
15	$\bar{2}$.96 143	$\bar{2}$.96 325	1.03 675	$\bar{1}$.99 817	$\bar{1}$.03 690	$\bar{1}$.03 948	0.96 052	$\bar{1}$.99 741	45
16	96 280	96 464	03 536	99 816	03 805	04 065	95 935	99 740	44
17	96 417	96 602	03 398	99 815	03 920	04 181	95 819	99 738	43
18	96 553	96 739	03 261	99 814	04 034	04 297	95 703	99 737	42
19	96 689	96 877	03 123	99 813	04 149	04 413	95 587	99 736	41
20	$\bar{2}$.96 825	$\bar{2}$.97 013	1.02 987	$\bar{1}$.99 812	$\bar{1}$.04 262	$\bar{1}$.04 528	0.95 472	$\bar{1}$.99 734	40
21	96 960	97 150	02 850	99 810	04 376	04 643	95 357	99 733	39
22	97 095	97 285	02 715	99 809	04 490	04 758	95 242	99 731	38
23	97 229	97 421	02 579	99 808	04 603	04 873	95 127	99 730	37
24	97 363	97 556	02 444	99 807	04 715	04 987	95 013	99 728	36
25	$\bar{2}$.97 496	$\bar{2}$.97 691	1.02 309	$\bar{1}$.99 806	$\bar{1}$.04 828	$\bar{1}$.05 101	0.94 899	$\bar{1}$.99 727	35
26	97 629	97 825	02 175	99 804	04 940	05 214	94 786	99 726	34
27	97 762	97 959	02 041	99 803	05 052	05 328	94 672	99 724	33
28	97 894	98 092	01 908	99 802	05 164	05 441	94 559	99 723	32
29	98 026	98 225	01 775	99 801	05 275	05 553	94 447	99 721	31
30	$\bar{2}$.98 157	$\bar{2}$.98 358	1.01 642	$\bar{1}$.99 800	$\bar{1}$.05 386	$\bar{1}$.05 666	0.94 334	$\bar{1}$.99 720	30
31	98 288	98 490	01 510	99 798	05 497	05 778	94 222	99 718	29
32	98 419	98 622	01 378	99 797	05 607	05 890	94 110	99 717	28
33	98 549	98 753	01 247	99 796	05 717	06 002	93 998	99 716	27
34	98 679	98 884	01 116	99 795	05 827	06 113	93 887	99 714	26
35	$\bar{2}$.98 808	$\bar{2}$.99 015	1.00 985	$\bar{1}$.99 793	$\bar{1}$.05 937	$\bar{1}$.06 224	0.93 776	$\bar{1}$.99 713	25
36	98 937	99 145	00 855	99 792	06 046	06 335	93 665	99 711	24
37	99 066	99 275	00 725	99 791	06 155	06 445	93 555	99 710	23
38	99 194	99 405	00 595	99 790	06 264	06 556	93 444	99 708	22
39	99 322	99 534	00 466	99 788	06 372	06 666	93 334	99 707	21
40	$\bar{2}$.99 450	$\bar{2}$.99 662	1.00 338	$\bar{1}$.99 787	$\bar{1}$.06 481	$\bar{1}$.06 775	0.93 225	$\bar{1}$.99 705	20
41	$\bar{2}$.99 577	$\bar{2}$.99 791	1.00 209	99 786	06 589	06 885	93 115	99 704	19
42	$\bar{2}$.99 704	$\bar{2}$.99 919	1.00 081	99 785	06 696	06 994	93 006	99 702	18
43	$\bar{2}$.99 830	$\bar{1}$.00 046	0.99 954	99 783	06 804	07 103	92 897	99 701	17
44	$\bar{2}$.99 956	$\bar{1}$.00 174	0.99 826	99 782	06 911	07 211	92 789	99 699	16
45	$\bar{1}$.00 082	$\bar{1}$.00 301	0.99 699	$\bar{1}$.99 781	$\bar{1}$.07 018	$\bar{1}$.07 320	0.92 680	$\bar{1}$.99 698	15
46	00 207	00 427	99 573	99 780	07 124	07 428	92 572	99 696	14
47	00 332	00 553	99 447	99 778	07 231	07 536	92 464	99 695	13
48	00 456	00 679	99 321	99 777	07 337	07 643	92 357	99 693	12
49	00 581	00 805	99 195	99 776	07 442	07 751	92 249	99 692	11
50	$\bar{1}$.00 704	$\bar{1}$.00 930	0.99 070	$\bar{1}$.99 775	$\bar{1}$.07 548	$\bar{1}$.07 858	0.92 142	$\bar{1}$.99 690	10
51	00 828	01 055	98 945	99 773	07 653	07 964	92 036	99 689	9
52	00 951	01 179	98 821	99 772	07 758	08 071	91 929	99 687	8
53	01 074	01 303	98 697	99 771	07 863	08 177	91 823	99 686	7
54	01 196	01 427	98 573	99 769	07 968	08 283	91 717	99 684	6
55	$\bar{1}$.01 318	$\bar{1}$.01 550	0.98 450	$\bar{1}$.99 768	$\bar{1}$.08 072	$\bar{1}$.08 389	0.91 611	$\bar{1}$.99 683	5
56	01 440	01 673	98 327	99 767	08 176	08 495	91 505	99 681	4
57	01 561	01 796	98 204	99 765	08 280	08 600	91 400	99 680	3
58	01 682	01 918	98 082	99 764	08 383	08 705	91 295	99 678	2
59	01 803	02 040	97 960	99 763	08 486	08 810	91 190	99 677	1
60′	$\bar{1}$.01 923	$\bar{1}$.02 162	0.97 838	$\bar{1}$.99 761	$\bar{1}$.08 589	$\bar{1}$.08 914	0.91 086	$\bar{1}$.99 675	0′
	log cos (84°)	log cot (84°)	log tan (84°)	log sin (84°)	log cos (83°)	log cot (83°)	log tan (83°)	log sin (83°)	′

	7°				8°				
′	log sin	log tan	log cot	log cos	log sin	log tan	log cot	log cos	
0′	1̄.08 589	1̄.08 914	0.91 086	1̄.99 675	1̄.14 356	1̄.14 780	0.85 220	1̄.99 575	60′
1	08 692	09 019	90 981	99 674	14 445	14 872	85 128	99 574	59
2	08 795	09 123	90 877	99 672	14 535	14 963	85 037	99 572	58
3	08 897	09 227	90 773	99 670	14 624	15 054	84 946	99 570	57
4	08 999	09 330	90 670	99 669	14 714	15 145	84 855	99 568	56
5	1̄.09 101	1̄.09 434	0.90 566	1̄.99 667	1̄.14 803	1̄.15 236	0.84 764	1̄.99 566	55
6	09 202	09 537	90 463	99 666	14 891	15 327	84 673	99 565	54
7	09 304	09 640	90 360	99 664	14 980	15 417	84 583	99 563	53
8	09 405	09 742	90 258	99 663	15 069	15 508	84 492	99 561	52
9	09 506	09 845	90 155	99 661	15 157	15 598	84 402	99 559	51
10	1̄.09 606	1̄.09 947	0.90 053	1̄.99 659	1̄.15 245	1̄.15 688	0.84 312	1̄.99 557	50
11	09 707	10 049	89 951	99 658	15 333	15 777	84 223	99 556	49
12	09 807	10 150	89 850	99 656	15 421	15 867	84 133	99 554	48
13	09 907	10 252	89 748	99 655	15 508	15 956	84 044	99 552	47
14	10 006	10 353	89 647	99 653	15 596	16 046	83 954	99 550	46
15	1̄.10 106	1̄.10 454	0.89 546	1̄.99 651	1̄.15 683	1̄.16 135	0.83 865	1̄.99 548	45
16	10 205	10 555	89 445	99 650	15 770	16 224	83 776	99 546	44
17	10 304	10 656	89 344	99 648	15 857	16 312	83 688	99 545	43
18	10 402	10 756	89 244	99 647	15 944	16 401	83 599	99 543	42
19	10 501	10 856	89 144	99 645	16 030	16 489	83 511	99 541	41
20	1̄.10 599	1̄.10 956	0.89 044	1̄.99 643	1̄.16 116	1̄.16 577	0.83 423	1̄.99 539	40
21	10 697	11 056	88 944	99 642	16 203	16 665	83 335	99 537	39
22	10 795	11 155	88 845	99 640	16 289	16 753	83 247	99 535	38
23	10 893	11 254	88 746	99 638	16 374	16 841	83 159	99 533	37
24	10 990	11 353	88 647	99 637	16 460	16 928	83 072	99 532	36
25	1̄.11 087	1̄.11 452	0.88 548	1̄.99 635	1̄.16 545	1̄.17 016	0.82 984	1̄.99 530	35
26	11 184	11 551	88 449	99 633	16 631	17 103	82 897	99 528	34
27	11 281	11 649	88 351	99 632	16 716	17 190	82 810	99 526	33
28	11 377	11 747	88 253	99 630	16 801	17 277	82 723	99 524	32
29	11 474	11 845	88 155	99 629	16 886	17 363	82 637	99 522	31
30	1̄.11 570	1̄.11 943	0.88 057	1̄.99 627	1̄.16 970	1̄.17 450	0.82 550	1̄.99 520	30
31	11 666	12 040	87 960	99 625	17 055	17 536	82 464	99 518	29
32	11 761	12 138	87 862	99 624	17 139	17 622	82 378	99 517	28
33	11 857	12 235	87 765	99 622	17 223	17 708	82 292	99 515	27
34	11 952	12 332	87 668	99 620	17 307	17 794	82 206	99 513	26
35	1̄.12 047	1̄.12 428	0.87 572	1̄.99 618	1̄.17 391	1̄.17 880	0.82 120	1̄.99 511	25
36	12 142	12 525	87 475	99 617	17 474	17 965	82 035	99 509	24
37	12 236	12 621	87 379	99 615	17 558	18 051	81 949	99 507	23
38	12 331	12 717	87 283	99 613	17 641	18 136	81 864	99 505	22
39	12 425	12 813	87 187	99 612	17 724	18 221	81 779	99 503	21
40	1̄.12 519	1̄.12 909	0.87 091	1̄.99 610	1̄.17 807	1̄.18 306	0.81 694	1̄.99 501	20
41	12 612	13 004	86 996	99 608	17 890	18 391	81 609	99 499	19
42	12 706	13 099	86 901	99 607	17 973	18 475	81 525	99 497	18
43	12 799	13 194	86 806	99 605	18 055	18 560	81 440	99 495	17
44	12 892	13 289	86 711	99 603	18 137	18 644	81 356	99 494	16
45	1̄.12 985	1̄.13 384	0.86 616	1̄.99 601	1̄.18 220	1̄.18 728	0.81 272	1̄.99 492	15
46	13 078	13 478	86 522	99 600	18 302	18 812	81 188	99 490	14
47	13 171	13 573	86 427	99 598	18 383	18 896	81 104	99 488	13
48	13 263	13 667	86 333	99 596	18 465	18 979	81 021	99 486	12
49	13 355	13 761	86 239	99 595	18 547	19 063	80 937	99 484	11
50	1̄.13 447	1̄.13 854	0.86 146	1̄.99 593	1̄.18 628	1̄.19 146	0.80 854	1̄.99 482	10
51	13 539	13 948	86 052	99 591	18 709	19 229	80 771	99 480	9
52	13 630	14 041	85 959	99 589	18 790	19 312	80 688	99 478	8
53	13 722	14 134	85 866	99 588	18 871	19 395	80 605	99 476	7
54	13 813	14 227	85 773	99 586	18 952	19 478	80 522	99 474	6
55	1̄.13 904	1̄.14 320	0.85 680	1̄.99 584	1̄.19 033	1̄.19 561	0.80 439	1̄.99 472	5
56	13 994	14 412	85 588	99 582	19 113	19 643	80 357	99 470	4
57	14 085	14 504	85 496	99 581	19 193	19 725	80 275	99 468	3
58	14 175	14 597	85 403	99 579	19 273	19 807	80 193	99 466	2
59	14 266	14 688	85 312	99 577	19 353	19 889	80 111	99 464	1
60′	1̄.14 356	1̄.14 780	0.85 220	1̄.99 575	1̄.19 433	1̄.19 971	0.80 029	1̄.99 462	0′
	log cos	log cot	log tan	log sin	log cos	log cot	log tan	log sin	′
	82°				81°				

	9°				10°				
′	log sin	log tan	log cot	log cos	log sin	log tan	log cot	log cos	
0′	$\bar{1}$.19 433	$\bar{1}$.19 971	0.80 029	$\bar{1}$.99 462	$\bar{1}$.23 967	$\bar{1}$.24 632	0.75 368	$\bar{1}$.99 335	60′
1	19 513	20 053	79 947	99 460	24 039	24 706	75 294	99 333	59
2	19 592	20 134	79 866	99 458	24 110	24 779	75 221	99 331	58
3	19 672	20 216	79 784	99 456	24 181	24 853	75 147	99 328	57
4	19 751	20 297	79 703	99 454	24 253	24 926	75 074	99 326	56
5	$\bar{1}$.19 830	$\bar{1}$.20 378	0.79 622	$\bar{1}$.99 452	$\bar{1}$.24 324	$\bar{1}$.25 000	0.75 000	$\bar{1}$.99 324	55
6	19 909	20 459	79 541	99 450	24 395	25 073	74 927	99 322	54
7	19 988	20 540	79 460	99 448	24 466	25 146	74 854	99 319	53
8	20 067	20 621	79 379	99 446	24 536	25 219	74 781	99 317	52
9	20 145	20 701	79 299	99 444	24 607	25 292	74 708	99 315	51
10	$\bar{1}$.20 223	$\bar{1}$.20 782	0.79 218	$\bar{1}$.99 442	$\bar{1}$.24 677	$\bar{1}$.25 365	0.74 635	$\bar{1}$.99 313	50
11	20 302	20 862	79 138	99 440	24 748	25 437	74 563	99 310	49
12	20 380	20 942	79 058	99 438	24 818	25 510	74 490	99 308	48
13	20 458	21 022	78 978	99 436	24 888	25 582	74 418	99 306	47
14	20 535	21 102	78 898	99 434	24 958	25 655	74 345	99 304	46
15	$\bar{1}$.20 613	$\bar{1}$.21 182	0.78 818	$\bar{1}$.99 432	$\bar{1}$.25 028	$\bar{1}$.25 727	0.74 273	$\bar{1}$.99 301	45
16	20 691	21 261	78 739	99 429	25 098	25 799	74 201	99 299	44
17	20 768	21 341	78 659	99 427	25 168	25 871	74 129	99 297	43
18	20 845	21 420	78 580	99 425	25 237	25 943	74 057	99 294	42
19	20 922	21 499	78 501	99 423	25 307	26 015	73 985	99 292	41
20	$\bar{1}$.20 999	$\bar{1}$.21 578	0.78 422	$\bar{1}$.99 421	$\bar{1}$.25 376	$\bar{1}$.26 086	0.73 914	$\bar{1}$.99 290	40
21	21 076	21 657	78 343	99 419	25 445	26 158	73 842	99 288	39
22	21 153	21 736	78 264	99 417	25 514	26 229	73 771	99 285	38
23	21 229	21 814	78 186	99 415	25 583	26 301	73 699	99 283	37
24	21 306	21 893	78 107	99 413	25 652	26 372	73 628	99 281	36
25	$\bar{1}$.21 382	$\bar{1}$.21 971	0.78 029	$\bar{1}$.99 411	$\bar{1}$.25 721	$\bar{1}$.26 443	0.73 557	$\bar{1}$.99 278	35
26	21 458	22 049	77 951	99 409	25 790	26 514	73 486	99 276	34
27	21 534	22 127	77 873	99 407	25 858	26 585	73 415	99 274	33
28	21 610	22 205	77 795	99 404	25 927	26 655	73 345	99 271	32
29	21 685	22 283	77 717	99 402	25 995	26 726	73 274	99 269	31
30	$\bar{1}$.21 761	$\bar{1}$.22 361	0.77 639	$\bar{1}$.99 400	$\bar{1}$.26 063	$\bar{1}$.26 797	0.73 203	$\bar{1}$.99 267	30
31	21 836	22 438	77 562	99 398	26 131	26 867	73 133	99 264	29
32	21 912	22 516	77 484	99 396	26 199	26 937	73 063	99 262	28
33	21 987	22 593	77 407	99 394	26 267	27 008	72 992	99 260	27
34	22 062	22 670	77 330	99 392	26 335	27 078	72 922	99 257	26
35	$\bar{1}$.22 137	$\bar{1}$.22 747	0.77 253	$\bar{1}$.99 390	$\bar{1}$.26 403	$\bar{1}$.27 148	0.72 852	$\bar{1}$.99 255	25
36	22 211	22 824	77 176	99 388	26 470	27 218	72 782	99 252	24
37	22 286	22 901	77 099	99 385	26 538	27 288	72 712	99 250	23
38	22 361	22 977	77 023	99 383	26 605	27 357	72 643	99 248	22
39	22 435	23 054	76 946	99 381	26 672	27 427	72 573	99 245	21
40	$\bar{1}$.22 509	$\bar{1}$.23 130	0.76 870	$\bar{1}$.99 379	$\bar{1}$.26 739	$\bar{1}$.27 496	0.72 504	$\bar{1}$.99 243	20
41	22 583	23 206	76 794	99 377	26 806	27 566	72 434	99 241	19
42	22 657	23 283	76 717	99 375	26 873	27 635	72 365	99 238	18
43	22 731	23 359	76 641	99 372	26 940	27 704	72 296	99 236	17
44	22 805	23 435	76 565	99 370	27 007	27 773	72 227	99 233	16
45	$\bar{1}$.22 878	$\bar{1}$.23 510	0.76 490	$\bar{1}$.99 368	$\bar{1}$.27 073	$\bar{1}$.27 842	0.72 158	$\bar{1}$.99 231	15
46	22 952	23 586	76 414	99 366	27 140	27 911	72 089	99 229	14
47	23 025	23 661	76 339	99 364	27 206	27 980	72 020	99 226	13
48	23 098	23 737	76 263	99 362	27 273	28 049	71 951	99 224	12
49	23 171	23 812	76 188	99 359	27 339	28 117	71 883	99 221	11
50	$\bar{1}$.23 244	$\bar{1}$.23 887	0.76 113	$\bar{1}$.99 357	$\bar{1}$.27 405	$\bar{1}$.28 186	0.71 814	$\bar{1}$.99 219	10
51	23 317	23 962	76 038	99 355	27 471	28 254	71 746	99 217	9
52	23 390	24 037	75 963	99 353	27 537	28 323	71 677	99 214	8
53	23 462	24 112	75 888	99 351	27 602	28 391	71 609	99 212	7
54	23 535	24 186	75 814	99 348	27 668	28 459	71 541	99 209	6
55	$\bar{1}$.23 607	$\bar{1}$.24 261	0.75 739	$\bar{1}$.99 346	$\bar{1}$.27 734	$\bar{1}$.28 527	0.71 473	$\bar{1}$.99 207	5
56	23 679	24 335	75 665	99 344	27 799	28 595	71 405	99 204	4
57	23 752	24 410	75 590	99 342	27 864	28 662	71 338	99 202	3
58	23 823	24 484	75 516	99 340	27 930	28 730	71 270	99 200	2
59	23 895	24 558	75 442	99 337	27 995	28 798	71 202	99 197	1
60′	$\bar{1}$.23 967	$\bar{1}$.24 632	0.75 368	$\bar{1}$.99 335	$\bar{1}$.28 060	$\bar{1}$.28 865	0.71 135	$\bar{1}$.99 195	0′
	log cos	log cot	log tan	log sin	log cos	log cot	log tan	log sin	′
	80°				79°				

′	11° log sin	log tan	log cot	log cos	12° log sin	log tan	log cot	log cos	
0′	$\bar{1}$.28 060	$\bar{1}$.28 865	0.71 135	$\bar{1}$.99 195	$\bar{1}$.31 788	$\bar{1}$.32 747	0.67 253	$\bar{1}$.99 040	60′
1	28 125	28 933	71 067	99 192	31 847	32 810	67 190	99 038	59
2	28 190	29 000	71 000	99 190	31 907	32 872	67 128	99 035	58
3	28 254	29 067	70 933	99 187	31 966	32 933	67 067	99 032	57
4	28 319	29 134	70 866	99 185	32 025	32 995	67 005	99 030	56
5	$\bar{1}$.28 384	$\bar{1}$.29 201	0.70 799	$\bar{1}$.99 182	$\bar{1}$.32 084	$\bar{1}$.33 057	0.66 943	$\bar{1}$.99 027	55
6	28 448	29 268	70 732	99 180	32 143	33 119	66 881	99 024	54
7	28 512	29 335	70 665	99 177	32 202	33 180	66 820	99 022	53
8	28 577	29 402	70 598	99 175	32 261	33 242	66 758	99 019	52
9	28 641	29 468	70 532	99 172	32 319	33 303	66 697	99 016	51
10	$\bar{1}$.28 705	$\bar{1}$.29 535	0.70 465	$\bar{1}$.99 170	$\bar{1}$.32 378	$\bar{1}$.33 365	0.66 635	$\bar{1}$.99 013	50
11	28 769	29 601	70 399	99 167	32 437	33 426	66 574	99 011	49
12	28 833	29 668	70 332	99 165	32 495	33 487	66 513	99 008	48
13	28 896	29 734	70 266	99 162	32 553	33 548	66 452	99 005	47
14	28 960	29 800	70 200	99 160	32 612	33 609	66 391	99 002	46
15	$\bar{1}$.29 024	$\bar{1}$.29 866	0.70 134	$\bar{1}$.99 157	$\bar{1}$.32 670	$\bar{1}$.33 670	0.66 330	$\bar{1}$.99 000	45
16	29 087	29 932	70 068	99 155	32 728	33 731	66 269	98 997	44
17	29 150	29 998	70 002	99 152	32 786	33 792	66 208	98 994	43
18	29 214	30 064	69 936	99 150	32 844	33 853	66 147	98 991	42
19	29 277	30 130	69 870	99 147	32 902	33 913	66 087	98 989	41
20	$\bar{1}$.29 340	$\bar{1}$.30 195	0.69 805	$\bar{1}$.99 145	$\bar{1}$.32 960	$\bar{1}$.33 974	0.66 026	$\bar{1}$.98 986	40
21	29 403	30 261	69 739	99 142	33 018	34 034	65 966	98 983	39
22	29 466	30 326	69 674	99 140	33 075	34 095	65 905	98 980	38
23	29 529	30 391	69 609	99 137	33 133	34 155	65 845	98 978	37
24	29 591	30 457	69 543	99 135	33 190	34 215	65 785	98 975	36
25	$\bar{1}$.29 654	$\bar{1}$.30 522	0.69 478	$\bar{1}$.99 132	$\bar{1}$.33 248	$\bar{1}$.34 276	0.65 724	$\bar{1}$.98 972	35
26	29 716	30 587	69 413	99 130	33 305	34 336	65 664	98 969	34
27	29 779	30 652	69 348	99 127	33 362	34 396	65 604	98 967	33
28	29 841	30 717	69 283	99 124	33 420	34 456	65 544	98 964	32
29	29 903	30 782	69 218	99 122	33 477	34 516	65 484	98 961	31
30	$\bar{1}$.29 966	$\bar{1}$.30 846	0.69 154	$\bar{1}$.99 119	$\bar{1}$.33 534	$\bar{1}$.34 576	0.65 424	$\bar{1}$.98 958	30
31	30 028	30 911	69 089	99 117	33 591	34 635	65 365	98 955	29
32	30 090	30 975	69 025	99 114	33 647	34 695	65 305	98 953	28
33	30 151	31 040	68 960	99 112	33 704	34 755	65 245	98 950	27
34	30 213	31 104	68 896	99 109	33 761	34 814	65 186	98 947	26
35	$\bar{1}$.30 275	$\bar{1}$.31 168	0.68 832	$\bar{1}$.99 106	$\bar{1}$.33 818	$\bar{1}$.34 874	0.65 126	$\bar{1}$.98 944	25
36	30 336	31 233	68 767	99 104	33 874	34 933	65 067	98 941	24
37	30 398	31 297	68 703	99 101	33 931	34 992	65 008	98 938	23
38	30 459	31 361	68 639	99 099	33 987	35 051	64 949	98 936	22
39	30 521	31 425	68 575	99 096	34 043	35 111	64 889	98 933	21
40	$\bar{1}$.30 582	$\bar{1}$.31 489	0.68 511	$\bar{1}$.99 093	$\bar{1}$.34 100	$\bar{1}$.35 170	0.64 830	$\bar{1}$.98 930	20
41	30 643	31 552	68 448	99 091	34 156	35 229	64 771	98 927	19
42	30 704	31 616	68 384	99 088	34 212	35 288	64 712	98 924	18
43	30 765	31 679	68 321	99 086	34 268	35 347	64 653	98 921	17
44	30 826	31 743	68 257	99 083	34 324	35 405	64 595	98 919	16
45	$\bar{1}$.30 887	$\bar{1}$.31 806	0.68 194	$\bar{1}$.99 080	$\bar{1}$.34 380	$\bar{1}$.35 464	0.64 536	$\bar{1}$.98 916	15
46	30 947	31 870	68 130	99 078	34 436	35 523	64 477	98 913	14
47	31 008	31 933	68 067	99 075	34 491	35 581	64 419	98 910	13
48	31 068	31 996	68 004	99 072	34 547	35 640	64 360	98 907	12
49	31 129	32 059	67 941	99 070	34 602	35 698	64 302	98 904	11
50	$\bar{1}$.31 189	$\bar{1}$.32 122	0.67 878	$\bar{1}$.99 067	$\bar{1}$.34 658	$\bar{1}$.35 757	0.64 243	$\bar{1}$.98 901	10
51	31 250	32 185	67 815	99 064	34 713	35 815	64 185	98 898	9
52	31 310	32 248	67 752	99 062	34 769	35 873	64 127	98 896	8
53	31 370	32 311	67 689	99 059	34 824	35 931	64 069	98 893	7
54	31 430	32 373	67 627	99 056	34 879	35 989	64 011	98 890	6
55	$\bar{1}$.31 490	$\bar{1}$.32 436	0.67 564	$\bar{1}$.99 054	$\bar{1}$.34 934	$\bar{1}$.36 047	0.63 953	$\bar{1}$.98 887	5
56	31 549	32 498	67 502	99 051	34 989	36 105	63 895	98 884	4
57	31 609	32 561	67 439	99 048	35 044	36 163	63 837	98 881	3
58	31 669	32 623	67 377	99 046	35 099	36 221	63 779	98 878	2
59	31 728	32 685	67 315	99 043	35 154	36 279	63 721	98 875	1
60′	$\bar{1}$.31 788	$\bar{1}$.32 747	0.67 253	$\bar{1}$.99 040	$\bar{1}$.35 209	$\bar{1}$.36 336	0.63 664	$\bar{1}$.98 872	0′
	log cos	log cot	log tan	log sin	log cos	log cot	log tan	log sin	′
	78°				77°				

′	13° log sin	log tan	log cot	log cos	14° log sin	log tan	log cot	log cos	
0′	$\bar{1}$.35 209	$\bar{1}$.36 336	0.63 664	$\bar{1}$.98 872	$\bar{1}$.38 368	$\bar{1}$.39 677	0.60 323	$\bar{1}$.98 690	60′
1	35 263	36 394	63 606	98 869	38 418	39 731	60 269	98 687	59
2	35 318	36 452	63 548	98 867	38 469	39 785	60 215	98 684	58
3	35 373	36 509	63 491	98 864	38 519	39 838	60 162	98 681	57
4	35 427	36 566	63 434	98 861	38 570	39 892	60 108	98 678	56
5	$\bar{1}$.35 481	$\bar{1}$.36 624	0.63 376	$\bar{1}$.98 858	$\bar{1}$.38 620	$\bar{1}$.39 945	0.60 055	$\bar{1}$.98 675	55
6	35 536	36 681	63 319	98 855	38 670	39 999	60 001	98 671	54
7	35 590	36 738	63 262	98 852	38 721	40 052	59 948	98 668	53
8	35 644	36 795	63 205	98 849	38 771	40 106	59 894	98 665	52
9	35 698	36 852	63 148	98 846	38 821	40 159	59 841	98 662	51
10	$\bar{1}$.35 752	$\bar{1}$.36 909	0.63 091	$\bar{1}$.98 843	$\bar{1}$.38 871	$\bar{1}$.40 212	0.59 788	$\bar{1}$.98 659	50
11	35 806	36 966	63 034	98 840	38 921	40 266	59 734	98 656	49
12	35 860	37 023	62 977	98 837	38 971	40 319	59 681	98 652	48
13	35 914	37 080	62 920	98 834	39 021	40 372	59 628	98 649	47
14	35 968	37 137	62 863	98 831	39 071	40 425	59 575	98 646	46
15	$\bar{1}$.36 022	$\bar{1}$.37 193	0.62 807	$\bar{1}$.98 828	$\bar{1}$.39 121	$\bar{1}$.40 478	0.59 522	$\bar{1}$.98 643	45
16	36 075	37 250	62 750	98 825	39 170	40 531	59 469	98 640	44
17	36 129	37 306	62 694	98 822	39 220	40 584	59 416	98 636	43
18	36 182	37 363	62 637	98 819	39 270	40 636	59 364	98 633	42
19	36 236	37 419	62 581	98 816	39 319	40 689	59 311	98 630	41
20	$\bar{1}$.36 289	$\bar{1}$.37 476	0.62 524	$\bar{1}$.98 813	$\bar{1}$.39 369	$\bar{1}$.40 742	0.59 258	$\bar{1}$.98 627	40
21	36 342	37 532	62 468	98 810	39 418	40 795	59 205	98 623	39
22	36 395	37 588	62 412	98 807	39 467	40 847	59 153	98 620	38
23	36 449	37 644	62 356	98 804	39 517	40 900	59 100	98 617	37
24	36 502	37 700	62 300	98 801	39 566	40 952	59 048	98 614	36
25	$\bar{1}$.36 555	$\bar{1}$.37 756	0.62 244	$\bar{1}$.98 798	$\bar{1}$.39 615	$\bar{1}$.41 005	0.58 995	$\bar{1}$.98 610	35
26	36 608	37 812	62 188	98 795	39 664	41 057	58 943	98 607	34
27	36 660	37 868	62 132	98 792	39 713	41 109	58 891	98 604	33
28	36 713	37 924	62 076	98 789	39 762	41 161	58 839	98 601	32
29	36 766	37 980	62 020	98 786	39 811	41 214	58 786	98 597	31
30	$\bar{1}$.36 819	$\bar{1}$.38 035	0.61 965	$\bar{1}$.98 783	$\bar{1}$.39 860	$\bar{1}$.41 266	0.58 734	$\bar{1}$.98 594	30
31	36 871	38 091	61 909	98 780	39 909	41 318	58 682	98 591	29
32	36 924	38 147	61 853	98 777	39 958	41 370	58 630	98 588	28
33	36 976	38 202	61 798	98 774	40 006	41 422	58 578	98 584	27
34	37 028	38 257	61 743	98 771	40 055	41 474	58 526	98 581	26
35	$\bar{1}$.37 081	$\bar{1}$.38 313	0.61 687	$\bar{1}$.98 768	$\bar{1}$.40 103	$\bar{1}$.41 526	0.58 474	$\bar{1}$.98 578	25
36	37 133	38 368	61 632	98 765	40 152	41 578	58 422	98 574	24
37	37 185	38 423	61 577	98 762	40 200	41 629	58 371	98 571	23
38	37 237	38 479	61 521	98 759	40 249	41 681	58 319	98 568	22
39	37 289	38 534	61 466	98 756	40 297	41 733	58 267	98 565	21
40	$\bar{1}$.37 341	$\bar{1}$.38 589	0.61 411	$\bar{1}$.98 753	$\bar{1}$.40 346	$\bar{1}$.41 784	0.58 216	$\bar{1}$.98 561	20
41	37 393	38 644	61 356	98 750	40 394	41 836	58 164	98 558	19
42	37 445	38 699	61 301	98 746	40 442	41 887	58 113	98 555	18
43	37 497	38 754	61 246	98 743	40 490	41 939	58 061	98 551	17
44	37 549	38 808	61 192	98 740	40 538	41 990	58 010	98 548	16
45	$\bar{1}$.37 600	$\bar{1}$.38 863	0.61 137	$\bar{1}$.98 737	$\bar{1}$.40 586	$\bar{1}$.42 041	0.57 959	$\bar{1}$.98 545	15
46	37 652	38 918	61 082	98 734	40 634	42 093	57 907	98 541	14
47	37 703	38 972	61 028	98 731	40 682	42 144	57 856	98 538	13
48	37 755	39 027	60 973	98 728	40 730	42 195	57 805	98 535	12
49	37 806	39 082	60 918	98 725	40 778	42 246	57 754	98 531	11
50	$\bar{1}$.37 858	$\bar{1}$.39 136	0.60 864	$\bar{1}$.98 722	$\bar{1}$.40 825	$\bar{1}$.42 297	0.57 703	$\bar{1}$.98 528	10
51	37 909	39 190	60 810	98 719	40 873	42 348	57 652	98 525	9
52	37 960	39 245	60 755	98 715	40 921	42 399	57 601	98 521	8
53	38 011	39 299	60 701	98 712	40 968	42 450	57 550	98 518	7
54	38 062	39 353	60 647	98 709	41 016	42 501	57 499	98 515	6
55	$\bar{1}$.38 113	$\bar{1}$.39 407	0.60 593	$\bar{1}$.98 706	$\bar{1}$.41 063	$\bar{1}$.42 552	0.57 448	$\bar{1}$.98 511	5
56	38 164	39 461	60 539	98 703	41 111	42 603	57 397	98 508	4
57	38 215	39 515	60 485	98 700	41 158	42 653	57 347	98 505	3
58	38 266	39 569	60 431	98 697	41 205	42 704	57 296	98 501	2
59	38 317	39 623	60 377	98 694	41 252	42 755	57 245	98 498	1
60	$\bar{1}$.38 368	$\bar{1}$.39 677	0.60 323	$\bar{1}$.98 690	$\bar{1}$.41 300	$\bar{1}$.42 805	0.57 195	$\bar{1}$.98 494	0′
	log cos	log cot	log tan	log sin 76°	log cos	log cot	log tan	log sin 75°	′

′	15° log sin	log tan	log cot	log cos	16° log sin	log tan	log cot	log cos	
0′	1̄.41 300	1̄.42 805	0.57 195	1̄.98 494	1̄.44 034	1̄.45 750	0.54 250	1̄.98 284	60′
1	41 347	42 856	57 144	98 491	44 078	45 797	54 203	98 281	59
2	41 394	42 906	57 094	98 488	44 122	45 845	54 155	98 277	58
3	41 441	42 957	57 043	98 484	44 166	45 892	54 108	98 273	57
4	41 488	43 007	56 993	98 481	44 210	45 940	54 060	98 270	56
5	1̄.41 535	1̄.43 057	0.56 943	1̄.98 477	1̄.44 253	1̄.45 987	0.54 013	1̄.98 266	55
6	41 582	43 108	56 892	98 474	44 297	46 035	53 965	98 262	54
7	41 628	43 158	56 842	98 471	44 341	46 082	53 918	98 259	53
8	41 675	43 208	56 792	98 467	44 385	46 130	53 870	98 255	52
9	41 722	43 258	56 742	98 464	44 428	46 177	53 823	98 251	51
10	1̄.41 768	1̄.43 308	0.56 692	1̄.98 460	1̄.44 472	1̄.46 224	0.53 776	1̄.98 248	50
11	41 815	43 358	56 642	98 457	44 516	46 271	53 729	98 244	49
12	41 861	43 408	56 592	98 453	44 559	46 319	53 681	98 240	48
13	41 908	43 458	56 542	98 450	44 602	46 366	53 634	98 237	47
14	41 954	43 508	56 492	98 447	44 646	46 413	53 587	98 233	46
15	1̄.42 001	1̄.43 558	0.56 442	1̄.98 443	1̄.44 689	1̄.46 460	0.53 540	1̄.98 229	45
16	42 047	43 607	56 393	98 440	44 733	46 507	53 493	98 226	44
17	42 093	43 657	56 343	98 436	44 776	46 554	53 446	98 222	43
18	42 140	43 707	56 293	98 433	44 819	46 601	53 399	98 218	42
19	42 186	43 756	56 244	98 429	44 862	46 648	53 352	98 215	41
20	1̄.42 232	1̄.43 806	0.56 194	1̄.98 426	1̄.44 905	1̄.46 694	0.53 306	1̄.98 211	40
21	42 278	43 855	56 145	98 422	44 948	46 741	53 259	98 207	39
22	42 324	43 905	56 095	98 419	44 992	46 788	53 212	98 204	38
23	42 370	43 954	56 046	98 415	45 035	46 835	53 165	98 200	37
24	42 416	44 004	55 996	98 412	45 077	46 881	53 119	98 196	36
25	1̄.42 461	1̄.44 053	0.55 947	1̄.98 409	1̄.45 120	1̄.46 928	0.53 072	1̄.98 192	35
26	42 507	44 102	55 898	98 405	45 163	46 975	53 025	98 189	34
27	42 553	44 151	55 849	98 402	45 206	47 021	52 979	98 185	33
28	42 599	44 201	55 799	98 398	45 249	47 068	52 932	98 181	32
29	42 644	44 250	55 750	98 395	45 292	47 114	52 886	98 177	31
30	1̄.42 690	1̄.44 299	0.55 701	1̄.98 391	1̄.45 334	1̄.47 160	0.52 840	1̄.98 174	30
31	42 735	44 348	55 652	98 388	45 377	47 207	52 793	98 170	29
32	42 781	44 397	55 603	98 384	45 419	47 253	52 747	98 166	28
33	42 826	44 446	55 554	98 381	45 462	47 299	52 701	98 162	27
34	42 872	44 495	55 505	98 377	45 504	47 346	52 654	98 159	26
35	1̄.42 917	1̄.44 544	0.55 456	1̄.98 373	1̄.45 547	1̄.47 392	0.52 608	1̄.98 155	25
36	42 962	44 592	55 408	98 370	45 589	47 438	52 562	98 151	24
37	43 008	44 641	55 359	98 366	45 632	47 484	52 516	98 147	23
38	43 053	44 690	55 310	98 363	45 674	47 530	52 470	98 144	22
39	43 098	44 738	55 262	98 359	45 716	47 576	52 424	98 140	21
40	1̄.43 143	1̄.44 787	0.55 213	1̄.98 356	1̄.45 758	1̄.47 622	0.52 378	1̄.98 136	20
41	43 188	44 836	55 164	98 352	45 801	47 668	52 332	98 132	19
42	43 233	44 884	55 116	98 349	45 843	47 714	52 286	98 129	18
43	43 278	44 933	55 067	98 345	45 885	47 760	52 240	98 125	17
44	43 323	44 981	55 019	98 342	45 927	47 806	52 194	98 121	16
45	1̄.43 367	1̄.45 029	0.54 971	1̄.98 338	1̄.45 969	1̄.47 852	0.52 148	1̄.98 117	15
46	43 412	45 078	54 922	98 334	46 011	47 897	52 103	98 113	14
47	43 457	45 126	54 874	98 331	46 053	47 943	52 057	98 110	13
48	43 502	45 174	54 826	98 327	46 095	47 989	52 011	98 106	12
49	43 546	45 222	54 778	98 324	46 136	48 035	51 965	98 102	11
50	1̄.43 591	1̄.45 271	0.54 729	1̄.98 320	1̄.46 178	1̄.48 080	0.51 920	1̄.98 098	10
51	43 635	45 319	54 681	98 317	46 220	48 126	51 874	98 094	9
52	43 680	45 367	54 633	98 313	46 262	48 171	51 829	98 090	8
53	43 724	45 415	54 585	98 309	46 303	48 217	51 783	98 087	7
54	43 769	45 463	54 537	98 306	46 345	48 262	51 738	98 083	6
55	1̄.43 813	1̄.45 511	0.54 489	1̄.98 302	1̄.46 386	1̄.48 307	0.51 693	1̄.98 079	5
56	43 857	45 559	54 441	98 299	46 428	48 353	51 647	98 075	4
57	43 901	45 606	54 394	98 295	46 469	48 398	51 602	98 071	3
58	43 946	45 654	54 346	98 291	46 511	48 443	51 557	98 067	2
59	43 990	45 702	54 298	98 288	46 552	48 489	51 511	98 063	1
60′	1̄.44 034	1̄.45 750	0.54 250	1̄.98 284	1̄.46 594	1̄.48 534	0.51 466	1̄.98 060	0′
	log cos	log cot	log tan	log sin	log cos	log cot	log tan	log sin	′
	74°				73°				

17°					18°				
′	log sin	log tan	log cot	log cos	log sin	log tan	log cot	log cos	
0′	$\bar{1}$.46 594	$\bar{1}$.48 534	0.51 466	$\bar{1}$.98 060	$\bar{1}$.48 998	$\bar{1}$.51 178	0.48 822	$\bar{1}$.97 821	60′
1	46 635	48 579	51 421	98 056	49 037	51 221	48 779	97 817	59
2	46 676	48 624	51 376	98 052	49 076	51 264	48 736	97 812	58
3	46 717	48 669	51 331	98 048	49 115	51 306	48 694	97 808	57
4	46 758	48 714	51 286	98 044	49 153	51 349	48 651	97 804	56
5	$\bar{1}$.46 800	$\bar{1}$.48 759	0.51 241	$\bar{1}$.98 040	$\bar{1}$.49 192	$\bar{1}$.51 392	0.48 608	$\bar{1}$.97 800	55
6	46 841	48 804	51 196	98 036	49 231	51 435	48 565	97 796	54
7	46 882	48 849	51 151	98 032	49 269	51 478	48 522	97 792	53
8	46 923	48 894	51 106	98 029	49 308	51 520	48 480	97 788	52
9	46 964	48 939	51 061	98 025	49 347	51 563	48 437	97 784	51
10	$\bar{1}$.47 005	$\bar{1}$.48 984	0.51 016	$\bar{1}$.98 021	$\bar{1}$.49 385	$\bar{1}$.51 606	0.48 394	$\bar{1}$.97 779	50
11	47 045	49 029	50 971	98 017	49 424	51 648	48 352	97 775	49
12	47 086	49 073	50 927	98 013	49 462	51 691	48 309	97 771	48
13	47 127	49 118	50 882	98 009	49 500	51 734	48 266	97 767	47
14	47 168	49 163	50 837	98 005	49 539	51 776	48 224	97 763	46
15	$\bar{1}$.47 209	$\bar{1}$.49 207	0.50 793	$\bar{1}$.98 001	$\bar{1}$.49 577	$\bar{1}$.51 819	0.48 181	$\bar{1}$.97 759	45
16	47 249	49 252	50 748	97 997	49 615	51 861	48 139	97 754	44
17	47 290	49 296	50 704	97 993	49 654	51 903	48 097	97 750	43
18	47 330	49 341	50 659	97 989	49 692	51 946	48 054	97 746	42
19	47 371	49 385	50 615	97 986	49 730	51 988	48 012	97 742	41
20	$\bar{1}$.47 411	$\bar{1}$.49 430	0.50 570	$\bar{1}$.97 982	$\bar{1}$.49 768	$\bar{1}$.52 031	0.47 969	$\bar{1}$.97 738	40
21	47 452	49 474	50 526	97 978	49 806	52 073	47 927	97 734	39
22	47 492	49 519	50 481	97 974	49 844	52 115	47 885	97 729	38
23	47 533	49 563	50 437	97 970	49 882	52 157	47 843	97 725	37
24	47 573	49 607	50 393	97 966	49 920	52 200	47 800	97 721	36
25	$\bar{1}$.47 613	$\bar{1}$.49 652	0.50 348	$\bar{1}$.97 962	$\bar{1}$.49 958	$\bar{1}$.52 242	0.47 758	$\bar{1}$.97 717	35
26	47 654	49 696	50 304	97 958	49 996	52 284	47 716	97 713	34
27	47 694	49 740	50 260	97 954	50 034	52 326	47 674	97 708	33
28	47 734	49 784	50 216	97 950	50 072	52 368	47 632	97 704	32
29	47 774	49 828	50 172	97 946	50 110	52 410	47 590	97 700	31
30	$\bar{1}$.47 814	$\bar{1}$.49 872	0.50 128	$\bar{1}$.97 942	$\bar{1}$.50 148	$\bar{1}$.52 452	0.47 548	$\bar{1}$.97 696	30
31	47 854	49 916	50 084	97 938	50 185	52 494	47 506	97 691	29
32	47 894	49 960	50 040	97 934	50 223	52 536	47 464	97 687	28
33	47 934	50 004	49 996	97 930	50 261	52 578	47 422	97 683	27
34	47 974	50 048	49 952	97 926	50 298	52 620	47 380	97 679	26
35	$\bar{1}$.48 014	$\bar{1}$.50 092	0.49 908	$\bar{1}$.97 922	$\bar{1}$.50 336	$\bar{1}$.52 661	0.47 339	$\bar{1}$.97 674	25
36	48 054	50 136	49 864	97 918	50 374	52 703	47 297	97 670	24
37	48 094	50 180	49 820	97 914	50 411	52 745	47 255	97 666	23
38	48 133	50 223	49 777	97 910	50 449	52 787	47 213	97 662	22
39	48 173	50 267	49 733	97 906	50 486	52 829	47 171	97 657	21
40	$\bar{1}$.48 213	$\bar{1}$.50 311	0.49 689	$\bar{1}$.97 902	$\bar{1}$.50 523	$\bar{1}$.52 870	0.47 130	$\bar{1}$.97 653	20
41	48 252	50 355	49 645	97 898	50 561	52 912	47 088	97 649	19
42	48 292	50 398	49 602	97 894	50 598	52 953	47 047	97 645	18
43	48 332	50 442	49 558	97 890	50 635	52 995	47 005	97 640	17
44	48 371	50 485	49 515	97 886	50 673	53 037	46 963	97 636	16
45	$\bar{1}$.48 411	$\bar{1}$.50 529	0.49 471	$\bar{1}$.97 882	$\bar{1}$.50 710	$\bar{1}$.53 078	0.46 922	$\bar{1}$.97 632	15
46	48 450	50 572	49 428	97 878	50 747	53 120	46 880	97 628	14
47	48 490	50 616	49 384	97 874	50 784	53 161	46 839	97 623	13
48	48 529	50 659	49 341	97 870	50 821	53 202	46 798	97 619	12
49	48 568	50 703	49 297	97 866	50 858	53 244	46 756	97 615	11
50	$\bar{1}$.48 607	$\bar{1}$.50 746	0.49 254	$\bar{1}$.97 861	$\bar{1}$.50 896	$\bar{1}$.53 285	0.46 715	$\bar{1}$.97 610	10
51	48 647	50 789	49 211	97 857	50 933	53 327	46 673	97 606	9
52	48 686	50 833	49 167	97 853	50 970	53 368	46 632	97 602	8
53	48 725	50 876	49 124	97 849	51 007	53 409	46 591	97 597	7
54	48 764	50 919	49 081	97 845	51 043	53 450	46 550	97 593	6
55	$\bar{1}$.48 803	$\bar{1}$.50 962	0.49 038	$\bar{1}$.97 841	$\bar{1}$.51 080	$\bar{1}$.53 492	0.46 508	$\bar{1}$.97 589	5
56	48 842	51 005	48 995	97 837	51 117	53 533	46 467	97 584	4
57	48 881	51 048	48 952	97 833	51 154	53 574	46 426	97 580	3
58	48 920	51 092	48 908	97 829	51 191	53 615	46 385	97 576	2
59	48 959	51 135	48 865	97 825	51 227	53 656	46 344	97 571	1
60′	$\bar{1}$.48 998	$\bar{1}$.51 178	0.48 822	$\bar{1}$.97 821	$\bar{1}$.51 264	$\bar{1}$.53 697	0.46 303	$\bar{1}$.97 567	0′
	log cos	log cot	log tan	log sin	log cos	log cot	log tan	log sin	′
	72°				71°				

′	19° log sin	log tan	log cot	log cos	20° log sin	log tan	log cot	log cos	
0′	1̄.51 264	1̄.53 697	0.46 303	1̄.97 567	1̄.53 405	1̄.56 107	0.43 893	1̄.97 299	60′
1	51 301	53 738	46 262	97 563	53 440	56 146	43 854	97 294	59
2	51 338	53 779	46 221	97 558	53 475	56 185	43 815	97 289	58
3	51 374	53 820	46 180	97 554	53 509	56 224	43 776	97 285	57
4	51 411	53 861	46 139	97 550	53 544	56 264	43 736	97 280	56
5	1̄.51 447	1̄.53 902	0.46 098	1̄.97 545	1̄.53 578	1̄.56 303	0.43 697	1̄.97 276	55
6	51 484	53 943	46 057	97 541	53 613	56 342	43 658	97 271	54
7	51 520	53 984	46 016	97 536	53 647	56 381	43 619	97 266	53
8	51 557	54 025	45 975	97 532	53 682	56 420	43 580	97 262	52
9	51 593	54 065	45 935	97 528	53 716	56 459	43 541	97 257	51
10	1̄.51 629	1̄.54 106	0.45 894	1̄.97 523	1̄.53 751	1̄.56 498	0.43 502	1̄.97 252	50
11	51 666	54 147	45 853	97 519	53 785	56 537	43 463	97 248	49
12	51 702	54 187	45 813	97 515	53 819	56 576	43 424	97 243	48
13	51 738	54 228	45 772	97 510	53 854	56 615	43 385	97 238	47
14	51 774	54 269	45 731	97 506	53 888	56 654	43 346	97 234	46
15	1̄.51 811	1̄.54 309	0.45 691	1̄.97 501	1̄.53 922	1̄.56 693	0.43 307	1̄.97 229	45
16	51 847	54 350	45 650	97 497	53 957	56 732	43 268	97 224	44
17	51 883	54 390	45 610	97 492	53 991	56 771	43 229	97 220	43
18	51 919	54 431	45 569	97 488	54 025	56 810	43 190	97 215	42
19	51 955	54 471	45 529	97 484	54 059	56 849	43 151	97 210	41
20	1̄.51 991	1̄.54 512	0.45 488	1̄.97 479	1̄.54 093	1̄.56 887	0.43 113	1̄.97 206	40
21	52 027	54 552	45 448	97 475	54 127	56 926	43 074	97 201	39
22	52 063	54 593	45 407	97 470	54 161	56 965	43 035	97 196	38
23	52 099	54 633	45 367	97 466	54 195	57 004	42 996	97 192	37
24	52 135	54 673	45 327	97 461	54 229	57 042	42 958	97 187	36
25	1̄.52 171	1̄.54 714	0.45 286	1̄.97 457	1̄.54 263	1̄.57 081	0.42 919	1̄.97 182	35
26	52 207	54 754	45 246	97 453	54 297	57 120	42 880	97 178	34
27	52 242	54 794	45 206	97 448	54 331	57 158	42 842	97 173	33
28	52 278	54 835	45 165	97 444	54 365	57 197	42 803	97 168	32
29	52 314	54 875	45 125	97 439	54 399	57 235	42 765	97 163	31
30	1̄.52 350	1̄.54 915	0.45 085	1̄.97 435	1̄.54 433	1̄.57 274	0.42 726	1̄.97 159	30
31	52 385	54 955	45 045	97 430	54 466	57 312	42 688	97 154	29
32	52 421	54 995	45 005	97 426	54 500	57 351	42 649	97 149	28
33	52 456	55 035	44 965	97 421	54 534	57 389	42 611	97 145	27
34	52 492	55 075	44 925	97 417	54 567	57 428	42 572	97 140	26
35	1̄.52 527	1̄.55 115	0.44 885	1̄.97 412	1̄.54 601	1̄.57 466	0.42 534	1̄.97 135	25
36	52 563	55 155	44 845	97 408	54 635	57 504	42 496	97 130	24
37	52 598	55 195	44 805	97 403	54 668	57 543	42 457	97 126	23
38	52 634	55 235	44 765	97 399	54 702	57 581	42 419	97 121	22
39	52 669	55 275	44 725	97 394	54 735	57 619	42 381	97 116	21
40	1̄.52 705	1̄.55 315	0.44 685	1̄.97 390	1̄.54 769	1̄.57 658	0.42 342	1̄.97 111	20
41	52 740	55 355	44 645	97 385	54 802	57 696	42 304	97 107	19
42	52 775	55 395	44 605	97 381	54 836	57 734	42 266	97 102	18
43	52 811	55 434	44 566	97 376	54 869	57 772	42 228	97 097	17
44	52 846	55 474	44 526	97 372	54 903	57 810	42 190	97 092	16
45	1̄.52 881	1̄.55 514	0.44 486	1̄.97 367	1̄.54 936	1̄.57 849	0.42 151	1̄.97 087	15
46	52 916	55 554	44 446	97 363	54 969	57 887	42 113	97 083	14
47	52 951	55 593	44 407	97 358	55 003	57 925	42 075	97 078	13
48	52 986	55 633	44 367	97 353	55 036	57 963	42 037	97 073	12
49	53 021	55 673	44 327	97 349	55 069	58 001	41 999	97 068	11
50	1̄.53 056	1̄.55 712	0.44 288	1̄.97 344	1̄.55 102	1̄.58 039	0.41 961	1̄.97 063	10
51	53 092	55 752	44 248	97 340	55 136	58 077	41 923	97 059	9
52	53 126	55 791	44 209	97 335	55 169	58 115	41 885	97 054	8
53	53 161	55 831	44 169	97 331	55 202	58 153	41 847	97 049	7
54	53 196	55 870	44 130	97 326	55 235	58 191	41 809	97 044	6
55	1̄.53 231	1̄.55 910	0.44 090	1̄.97 322	1̄.55 268	1̄.58 229	0.41 771	1̄.97 039	5
56	53 266	55 949	44 051	97 317	55 301	58 267	41 733	97 035	4
57	53 301	55 989	44 011	97 312	55 334	58 304	41 696	97 030	3
58	53 336	56 028	43 972	97 308	55 367	58 342	41 658	97 025	2
59	53 370	56 067	43 933	97 303	55 400	58 380	41 620	97 020	1
60	1̄.53 405	1̄.56 107	0.43 893	1̄.97 299	1̄.55 433	1̄.58 418	0.41 582	1̄.97 015	0′
	log cos	log cot	log tan	log sin	log cos	log cot	log tan	log sin	′
	70°				69°				

	21°				22°				
′	log sin	log tan	log cot	log cos	log sin	log tan	log cot	log cos	
0′	$\bar{1}$.55 433	$\bar{1}$.58 418	0.41 582	$\bar{1}$.97 015	$\bar{1}$.57 358	$\bar{1}$.60 641	0.39 359	$\bar{1}$.96 717	**60′**
1	55 466	58 455	41 545	97 010	57 389	60 677	39 323	96 711	59
2	55 499	58 493	41 507	97 005	57 420	60 714	39 286	96 706	58
3	55 532	58 531	41 469	97 001	57 451	60 750	39 250	96 701	57
4	55 564	58 569	41 431	96 996	57 482	60 786	39 214	96 696	56
5	$\bar{1}$.55 597	$\bar{1}$.58 606	0.41 394	$\bar{1}$.96 991	$\bar{1}$.57 514	$\bar{1}$.60 823	0.39 177	$\bar{1}$.96 691	55
6	55 630	58 644	41 356	96 986	57 545	60 859	39 141	96 686	54
7	55 663	58 681	41 319	96 981	57 576	60 895	39 105	96 681	53
8	55 695	58 719	41 281	96 976	57 607	60 931	39 069	96 676	52
9	55 728	58 757	41 243	96 971	57 638	60 967	39 033	96 670	51
10	$\bar{1}$.55 761	$\bar{1}$.58 794	0.41 206	$\bar{1}$.96 966	$\bar{1}$.57 669	$\bar{1}$.61 004	0.38 996	$\bar{1}$.96 665	**50**
11	55 793	58 832	41 168	96 962	57 700	61 040	38 960	96 660	49
12	55 826	58 869	41 131	96 957	57 731	61 076	38 924	96 655	48
13	55 858	58 907	41 093	96 952	57 762	61 112	38 888	96 650	47
14	55 891	58 944	41 056	96 947	57 793	61 148	38 852	96 645	46
15	$\bar{1}$.55 923	$\bar{1}$.58 981	0.41 019	$\bar{1}$.96 942	$\bar{1}$.57 824	$\bar{1}$.61 184	0.38 816	$\bar{1}$.96 640	45
16	55 956	59 019	40 981	96 937	57 855	61 220	38 780	96 634	44
17	55 988	59 056	40 944	96 932	57 885	61 256	38 744	96 629	43
18	56 021	59 094	40 906	96 927	57 916	61 292	38 708	96 624	42
19	56 053	59 131	40 869	96 922	57 947	61 328	38 672	96 619	41
20	$\bar{1}$.56 085	$\bar{1}$.59 168	0.40 832	$\bar{1}$.96 917	$\bar{1}$.57 978	$\bar{1}$.61 364	0.38 636	$\bar{1}$.96 614	**40**
21	56 118	59 205	40 795	96 912	58 008	61 400	38 600	96 608	39
22	56 150	59 243	40 757	96 907	58 039	61 436	38 564	96 603	38
23	56 182	59 280	40 720	96 903	58 070	61 472	38 528	96 598	37
24	56 215	59 317	40 683	96 898	58 101	61 508	38 492	96 593	36
25	$\bar{1}$.56 247	$\bar{1}$.59 354	0.40 646	$\bar{1}$.96 893	$\bar{1}$.58 131	$\bar{1}$.61 544	0.38 456	$\bar{1}$.96 588	35
26	56 279	59 391	40 609	96 888	58 162	61 579	38 421	96 582	34
27	56 311	59 429	40 571	96 883	58 192	61 615	38 385	96 577	33
28	56 343	59 466	40 534	96 878	58 223	61 651	38 349	96 572	32
29	56 375	59 503	40 497	96 873	58 253	61 687	38 313	96 567	31
30	$\bar{1}$.56 408	$\bar{1}$.59 540	0.40 460	$\bar{1}$.96 868	$\bar{1}$.58 284	$\bar{1}$.61 722	0.38 278	$\bar{1}$.96 562	**30**
31	56 440	59 577	40 423	96 863	58 314	61 758	38 242	96 556	29
32	56 472	59 614	40 386	96 858	58 345	61 794	38 206	96 551	28
33	56 504	59 651	40 349	96 853	58 375	61 830	38 170	96 546	27
34	56 536	59 688	40 312	96 848	58 406	61 865	38 135	96 541	26
35	$\bar{1}$.56 568	$\bar{1}$.59 725	0.40 275	$\bar{1}$.96 843	$\bar{1}$.58 436	$\bar{1}$.61 901	0.38 099	$\bar{1}$.96 535	25
36	56 599	59 762	40 238	96 838	58 467	61 936	38 064	96 530	24
37	56 631	59 799	40 201	96 833	58 497	61 972	38 028	96 525	23
38	56 663	59 835	40 165	96 828	58 527	62 008	37 992	96 520	22
39	56 695	59 872	40 128	96 823	58 557	62 043	37 957	96 514	21
40	$\bar{1}$.56 727	$\bar{1}$.59 909	0.40 091	$\bar{1}$.96 818	$\bar{1}$.58 588	$\bar{1}$.62 079	0.37 921	$\bar{1}$.96 509	**20**
41	56 759	59 946	40 054	96 813	58 618	62 114	37 886	96 504	19
42	56 790	59 983	40 017	96 808	58 648	62 150	37 850	96 498	18
43	56 822	60 019	39 981	96 803	58 678	62 185	37 815	96 493	17
44	56 854	60 056	39 944	96 798	58 709	62 221	37 779	96 488	16
45	$\bar{1}$.56 886	$\bar{1}$.60 093	0.39 907	$\bar{1}$.96 793	$\bar{1}$.58 739	$\bar{1}$.62 256	0.37 744	$\bar{1}$.96 483	15
46	56 917	60 130	39 870	96 788	58 769	62 292	37 708	96 477	14
47	56 949	60 166	39 834	96 783	58 799	62 327	37 673	96 472	13
48	56 980	60 203	39 797	96 778	58 829	62 362	37 638	96 467	12
49	57 012	60 240	39 760	96 772	58 859	62 398	37 602	96 461	11
50	$\bar{1}$.57 044	$\bar{1}$.60 276	0.39 724	$\bar{1}$.96 767	$\bar{1}$.58 889	$\bar{1}$.62 433	0.37 567	$\bar{1}$.96 456	**10**
51	57 075	60 313	39 687	96 762	58 919	62 468	37 532	96 451	9
52	57 107	60 349	39 651	96 757	58 949	62 504	37 496	96 445	8
53	57 138	60 386	39 614	96 752	58 979	62 539	37 461	96 440	7
54	57 169	60 422	39 578	96 747	59 009	62 574	37 426	96 435	6
55	$\bar{1}$.57 201	$\bar{1}$.60 459	0.39 541	$\bar{1}$.96 742	$\bar{1}$.59 039	$\bar{1}$.62 609	0.37 391	$\bar{1}$.96 429	5
56	57 232	60 495	39 505	96 737	59 069	62 645	37 355	96 424	4
57	57 264	60 532	39 468	96 732	59 098	62 680	37 320	96 419	3
58	57 295	60 568	39 432	96 727	59 128	62 715	37 285	96 413	2
59	57 326	60 605	39 395	96 722	59 158	62 750	37 250	96 408	1
60′	$\bar{1}$.57 358	$\bar{1}$.60 641	0.39 359	$\bar{1}$.96 717	$\bar{1}$.59 188	$\bar{1}$.62 785	0.37 215	$\bar{1}$.96 403	**0′**
	log cos	log cot	log tan	log sin	log cos	log cot	log tan	log sin	′
	68°				67°				

′	23° log sin	log tan	log cot	log cos	24° log sin	log tan	log cot	log cos	
0′	1̄.59 188	1̄.62 785	0.37 215	1̄.96 403	1̄.60 931	1̄.64 858	0.35 142	1̄.96 073	60′
1	59 218	62 820	37 180	96 397	60 960	64 892	35 108	96 067	59
2	59 247	62 855	37 145	96 392	60 988	64 926	35 074	96 062	58
3	59 277	62 890	37 110	96 387	61 016	64 960	35 040	96 056	57
4	59 307	62 926	37 074	96 381	61 045	64 994	35 006	96 050	56
5	1̄.59 336	1̄.62 961	0.37 039	1̄.96 376	1̄.61 073	1̄.65 028	0.34 972	1̄.96 045	55
6	59 366	62 996	37 004	96 370	61 101	65 062	34 938	96 039	54
7	59 396	63 031	36 969	96 365	61 129	65 096	34 904	96 034	53
8	59 425	63 066	36 934	96 360	61 158	65 130	34 870	96 028	52
9	59 455	63 101	36 899	96 354	61 186	65 164	34 836	96 022	51
10	1̄.59 484	1̄.63 135	0.36 865	1̄.96 349	1̄.61 214	1̄.65 197	0.34 803	1̄.96 017	50
11	59 514	63 170	36 830	96 343	61 242	65 231	34 769	96 011	49
12	59 543	63 205	36 795	96 338	61 270	65 265	34 735	96 005	48
13	59 573	63 240	36 760	96 333	61 298	65 299	34 701	96 000	47
14	59 602	63 275	36 725	96 327	61 326	65 333	34 667	95 994	46
15	1̄.59 632	1̄.63 310	0.36 690	1̄.96 322	1̄.61 354	1̄.65 366	0.34 634	1̄.95 988	45
16	59 661	63 345	36 655	96 316	61 382	65 400	34 600	95 982	44
17	59 690	63 379	36 621	96 311	61 411	65 434	34 566	95 977	43
18	59 720	63 414	36 586	96 305	61 438	65 467	34 533	95 971	42
19	59 749	63 449	36 551	96 300	61 466	65 501	34 499	95 965	41
20	1̄.59 778	1̄.63 484	0.36 516	1̄.96 294	1̄.61 494	1̄.65 535	0.34 465	1̄.95 960	40
21	59 808	63 519	36 481	96 289	61 522	65 568	34 432	95 954	39
22	59 837	63 553	36 447	96 284	61 550	65 602	34 398	95 948	38
23	59 866	63 588	36 412	96 278	61 578	65 636	34 364	95 942	37
24	59 895	63 623	36 377	96 273	61 606	65 669	34 331	95 937	36
25	1̄.59 924	1̄.63 657	0.36 343	1̄.96 267	1̄.61 634	1̄.65 703	0.34 297	1̄.95 931	35
26	59 954	63 692	36 308	96 262	61 662	65 736	34 264	95 925	34
27	59 983	63 726	36 274	96 256	61 689	65 770	34 230	95 920	33
28	60 012	63 761	36 239	96 251	61 717	65 803	34 197	95 914	32
29	60 041	63 796	36 204	96 245	61 745	65 837	34 163	95 908	31
30	1̄.60 070	1̄.63 830	0.36 170	1̄.96 240	1̄.61 773	1̄.65 870	0.34 130	1̄.95 902	30
31	60 099	63 865	36 135	96 234	61 800	65 904	34 096	95 897	29
32	60 128	63 899	36 101	96 229	61 828	65 937	34 063	95 891	28
33	60 157	63 934	36 066	96 223	61 856	65 971	34 029	95 885	27
34	60 186	63 968	36 032	96 218	61 883	66 004	33 996	95 879	26
35	1̄.60 215	1̄.64 003	0.35 997	1̄.96 212	1̄.61 911	1̄.66 038	0.33 962	1̄.95 873	25
36	60 244	64 037	35 963	96 207	61 939	66 071	33 929	95 868	24
37	60 273	64 072	35 928	96 201	61 966	66 104	33 896	95 862	23
38	60 302	64 106	35 894	96 196	61 994	66 138	33 862	95 856	22
39	60 331	64 140	35 860	96 190	62 021	66 171	33 829	95 850	21
40	1̄.60 359	1̄.64 175	0.35 825	1̄.96 185	1̄.62 049	1̄.66 204	0.33 796	1̄.95 844	20
41	60 388	64 209	35 791	96 179	62 076	66 238	33 762	95 839	19
42	60 417	64 243	35 757	96 174	62 104	66 271	33 729	95 833	18
43	60 446	64 278	35 722	96 168	62 131	66 304	33 696	95 827	17
44	60 474	64 312	35 688	96 162	62 159	66 337	33 663	95 821	16
45	1̄.60 503	1̄.64 346	0.35 654	1̄.96 157	1̄.62 186	1̄.66 371	0.33 629	1̄.95 815	15
46	60 532	64 381	35 619	96 151	62 214	66 404	33 596	95 810	14
47	60 561	64 415	35 585	96 146	62 241	66 437	33 563	95 804	13
48	60 589	64 449	35 551	96 140	62 268	66 470	33 530	95 798	12
49	60 618	64 483	35 517	96 135	62 296	66 503	33 497	95 792	11
50	1̄.60 646	1̄.64 517	0.35 483	1̄.96 129	1̄.62 323	1̄.66 537	0.33 463	1̄.95 786	10
51	60 675	64 552	35 448	96 123	62 350	66 570	33 430	95 780	9
52	60 704	64 586	35 414	96 118	62 377	66 603	33 397	95 775	8
53	60 732	64 620	35 380	96 112	62 405	66 636	33 364	95 769	7
54	60 761	64 654	35 346	96 107	62 432	66 669	33 331	95 763	6
55	1̄.60 789	1̄.64 688	0.35 312	1̄.96 101	1̄.62 459	1̄.66 702	0.33 298	1̄.95 757	5
56	60 818	64 722	35 278	96 095	62 486	66 735	33 265	95 751	4
57	60 846	64 756	35 244	96 090	62 513	66 768	33 232	95 745	3
58	60 875	64 790	35 210	96 084	62 541	66 801	33 199	95 739	2
59	60 903	64 824	35 176	96 079	62 568	66 834	33 166	95 733	1
60′	1̄.60 931	1̄.64 858	0.35 142	1̄.96 073	1̄.62 595	1̄.66 867	0.33 133	1̄.95 728	0′
	log cos	log cot	log tan	log sin 66°	log cos	log cot	log tan	log sin 65°	′

	25				26°				
′	log sin	log tan	log cot	log cos	log sin	log tan	log cot	log cos	
0′	1̄.62 595	1̄.66 867	0.33 133	1̄.95 728	1̄.64 184	1̄.68 818	0.31 182	1̄.95 366	60′
1	62 622	66 900	33 100	95 722	64 210	68 850	31 150	95 360	59
2	62 649	66 933	33 067	95 716	64 236	68 882	31 118	95 354	58
3	62 676	66 966	33 034	95 710	64 262	68 914	31 086	95 348	57
4	62 703	66 999	33 001	95 704	64 288	68 946	31 054	95 341	56
5	1̄.62 730	1̄.67 032	0.32 968	1̄.95 698	1̄.64 313	1̄.68 978	0.31 022	1̄.95 335	55
6	62 757	67 065	32 935	95 692	64 339	69 010	30 990	95 329	54
7	62 784	67 098	32 902	95 686	64 365	69 042	30 958	95 323	53
8	62 811	67 131	32 869	95 680	64 391	69 074	30 926	95 317	52
9	62 838	67 163	32 837	95 674	64 417	69 106	30 894	95 310	51
10	1̄.62 865	1̄.67 196	0.32 804	1̄.95 668	1̄.64 442	1̄.69 138	0.30 862	1̄.95 304	50
11	62 892	67 229	32 771	95 663	64 468	69 170	30 830	95 298	49
12	62 918	67 262	32 738	95 657	64 494	69 202	30 798	95 292	48
13	62 945	67 295	32 705	95 651	64 519	69 234	30 766	95 286	47
14	62 972	67 327	32 673	95 645	64 545	69 266	30 734	95 279	46
15	1̄.62 999	1̄.67 360	0.32 640	1̄.95 639	1̄.64 571	1̄.69 298	0.30 702	1̄.95 273	45
16	63 026	67 393	32 607	95 633	64 596	69 329	30 671	95 267	44
17	63 052	67 426	32 574	95 627	64 622	69 361	30 639	95 261	43
18	63 079	67 458	32 542	95 621	64 647	69 393	30 607	95 254	42
19	63 106	67 491	32 509	95 615	64 673	69 425	30 575	95 248	41
20	1̄.63 133	1̄.67 524	0.32 476	1̄.95 609	1̄.64 698	1̄.69 457	0.30 543	1̄.95 242	40
21	63 159	67 556	32 444	95 603	64 724	69 488	30 512	95 236	39
22	63 186	67 589	32 411	95 597	64 749	69 520	30 480	95 229	38
23	63 213	67 622	32 378	95 591	64 775	69 552	30 448	95 223	37
24	63 239	67 654	32 346	95 585	64 800	69 584	30 416	95 217	36
25	1̄.63 266	1̄.67 687	0.32 313	1̄.95 579	1̄.64 826	1̄.69 615	0.30 385	1̄.95 211	35
26	63 292	67 719	32 281	95 573	64 851	69 647	30 353	95 204	34
27	63 319	67 752	32 248	95 567	64 877	69 679	30 321	95 198	33
28	63 345	67 785	32 215	95 561	64 902	69 710	30 290	95 192	32
29	63 372	67 817	32 183	95 555	64 927	69 742	30 258	95 185	31
30	1̄.63 398	1̄.67 850	0.32 150	1̄.95 549	1̄.64 953	1̄.69 774	0.30 226	1̄.95 179	30
31	63 425	67 882	32 118	95 543	64 978	69 805	30 195	95 173	29
32	63 451	67 915	32 085	95 537	65 003	69 837	30 163	95 167	28
33	63 478	67 947	32 053	95 531	65 029	69 868	30 132	95 160	27
34	63 504	67 980	32 020	95 525	65 054	69 900	30 100	95 154	26
35	1̄.63 531	1̄.68 012	0.31 988	1̄.95 519	1̄.65 079	1̄.69 932	0.30 068	1̄.95 148	25
36	63 557	68 044	31 956	95 513	65 104	69 963	30 037	95 141	24
37	63 583	68 077	31 923	95 507	65 130	69 995	30 005	95 135	23
38	63 610	68 109	31 891	95 500	65 155	70 026	29 974	95 129	22
39	63 636	68 142	31 858	95 494	65 180	70 058	29 942	95 122	21
40	1̄.63 662	1̄.68 174	0.31 826	1̄.95 488	1̄.65 205	1̄.70 089	0.29 911	1̄.95 116	20
41	63 689	68 206	31 794	95 482	65 230	70 121	29 879	95 110	19
42	63 715	68 239	31 761	95 476	65 255	70 152	29 848	95 103	18
43	63 741	68 271	31 729	95 470	65 281	70 184	29 816	95 097	17
44	63 767	68 303	31 697	95 464	65 306	70 215	29 785	95 090	16
45	1̄.63 794	1̄.68 336	0.31 664	1̄.95 458	1̄.65 331	1̄.70 247	0.29 753	1̄.95 084	15
46	63 820	68 368	31 632	95 452	65 356	70 278	29 722	95 078	14
47	63 846	68 400	31 600	95 446	65 381	70 309	29 691	95 071	13
48	63 872	68 432	31 568	95 440	65 406	70 341	29 659	95 065	12
49	63 898	68 465	31 535	95 434	65 431	70 372	29 628	95 059	11
50	1̄.63 924	1̄.68 497	0.31 503	1̄.95 427	1̄.65 456	1̄.70 404	0.29 596	1̄.95 052	10
51	63 950	68 529	31 471	95 421	65 481	70 435	29 565	95 046	9
52	63 976	68 561	31 439	95 415	65 506	70 466	29 534	95 039	8
53	64 002	68 593	31 407	95 409	65 531	70 498	29 502	95 033	7
54	64 028	68 626	31 374	95 403	65 556	70 529	29 471	95 027	6
55	1̄.64 054	1̄.68 658	0.31 342	1̄.95 397	1̄.65 580	1̄.70 560	0.29 440	1̄.95 020	5
56	64 080	68 690	31 310	95 391	65 605	70 592	29 408	95 014	4
57	64 106	68 722	31 278	95 384	65 630	70 623	29 377	95 007	3
58	64 132	68 754	31 246	95 378	65 655	70 654	29 346	95 001	2
59	64 158	68 786	31 214	95 372	65 680	70 685	29 315	94 995	1
60′	1̄.64 184	1̄.68 818	0.31 182	1̄.95 366	1̄.65 705	1̄.70 717	0.29 283	1̄.94 988	0′
	log cos	log cot	log tan	log sin	log cos	log cot	log tan	log sin	′
	64°				63°				

′	27° log sin	log tan	log cot	log cos	28° log sin	log tan	log cot	log cos	
0′	$\bar{1}$.65 705	$\bar{1}$.70 717	0.29 283	$\bar{1}$.94 988	$\bar{1}$.67 161	$\bar{1}$.72 567	0.27 433	$\bar{1}$.94 593	60′
1	65 729	70 748	29 252	94 982	67 185	72 598	27 402	94 587	59
2	65 754	70 779	29 221	94 975	67 208	72 628	27 372	94 580	58
3	65 779	70 810	29 190	94 969	67 232	72 659	27 341	94 573	57
4	65 804	70 841	29 159	94 962	67 256	72 689	27 311	94 567	56
5	$\bar{1}$.65 828	$\bar{1}$.70 873	0.29 127	$\bar{1}$.94 956	$\bar{1}$.67 280	$\bar{1}$.72 720	0.27 280	$\bar{1}$.94 560	55
6	65 853	70 904	29 096	94 949	67 303	72 750	27 250	94 553	54
7	65 878	70 935	29 065	94 943	67 327	72 780	27 220	94 546	53
8	65 902	70 966	29 034	94 936	67 350	72 811	27 189	94 540	52
9	65 927	70 997	29 003	94 930	67 374	72 841	27 159	94 533	51
10	$\bar{1}$.65 952	$\bar{1}$.71 028	0.28 972	$\bar{1}$.94 923	$\bar{1}$.67 398	$\bar{1}$.72 872	0.27 128	$\bar{1}$.94 526	50
11	65 976	71 059	28 941	94 917	67 421	72 902	27 098	94 519	49
12	66 001	71 090	28 910	94 911	67 445	72 932	27 068	94 513	48
13	66 025	71 121	28 879	94 904	67 468	72 963	27 037	94 506	47
14	66 050	71 153	28 847	94 898	67 492	72 993	27 007	94 499	46
15	$\bar{1}$.66 075	$\bar{1}$.71 184	0.28 816	$\bar{1}$.94 891	$\bar{1}$.67 515	$\bar{1}$.73 023	0.26 977	$\bar{1}$.94 492	45
16	66 099	71 215	28 785	94 885	67 539	73 054	26 946	94 485	44
17	66 124	71 246	28 754	94 878	67 562	73 084	26 916	94 479	43
18	66 148	71 277	28 723	94 871	67 586	73 114	26 886	94 472	42
19	66 173	71 308	28 692	94 865	67 609	73 144	26 856	94 465	41
20	$\bar{1}$.66 197	$\bar{1}$.71 339	0.28 661	$\bar{1}$.94 858	$\bar{1}$.67 633	$\bar{1}$.73 175	0.26 825	$\bar{1}$.94 458	40
21	66 221	71 370	28 630	94 852	67 656	73 205	26 795	94 451	39
22	66 246	71 401	28 599	94 845	67 680	73 235	26 765	94 445	38
23	66 270	71 431	28 569	94 839	67 703	73 265	26 735	94 438	37
24	66 295	71 462	28 538	94 832	67 726	73 295	26 705	94 431	36
25	$\bar{1}$.66 319	$\bar{1}$.71 493	0.28 507	$\bar{1}$.94 826	$\bar{1}$.67 750	$\bar{1}$.73 326	0.26 674	$\bar{1}$.94 424	35
26	66 343	71 524	28 476	94 819	67 773	73 356	26 644	94 417	34
27	66 368	71 555	28 445	94 813	67 796	73 386	26 614	94 410	33
28	66 392	71 586	28 414	94 806	67 820	73 416	26 584	94 404	32
29	66 416	71 617	28 383	94 799	67 843	73 446	26 554	94 397	31
30	$\bar{1}$.66 441	$\bar{1}$.71 648	0.28 352	$\bar{1}$.94 793	$\bar{1}$.67 866	$\bar{1}$.73 476	0.26 524	$\bar{1}$.94 390	30
31	66 465	71 679	28 321	94 786	67 890	73 507	26 493	94 383	29
32	66 489	71 709	28 291	94 780	67 913	73 537	26 463	94 376	28
33	66 513	71 740	28 260	94 773	67 936	73 567	26 433	94 369	27
34	66 537	71 771	28 229	94 767	67 959	73 597	26 403	94 362	26
35	$\bar{1}$.66 562	$\bar{1}$.71 802	0.28 198	$\bar{1}$.94 760	$\bar{1}$.67 982	$\bar{1}$.73 627	0.26 373	$\bar{1}$.94 355	25
36	66 586	71 833	28 167	94 753	68 006	73 657	26 343	94 349	24
37	66 610	71 863	28 137	94 747	68 029	73 687	26 313	94 342	23
38	66 634	71 894	28 106	94 740	68 052	73 717	26 283	94 335	22
39	66 658	71 925	28 075	94 734	68 075	73 747	26 253	94 328	21
40	$\bar{1}$.66 682	$\bar{1}$.71 955	0.28 045	$\bar{1}$.94 727	$\bar{1}$.68 098	$\bar{1}$.73 777	0.26 223	$\bar{1}$.94 321	20
41	66 706	71 986	28 014	94 720	68 121	73 807	26 193	94 314	19
42	66 731	72 017	27 983	94 714	68 144	73 837	26 163	94 307	18
43	66 755	72 048	27 952	94 707	68 167	73 867	26 133	94 300	17
44	66 779	72 078	27 922	94 700	68 190	73 897	26 103	94 293	16
45	$\bar{1}$.66 803	$\bar{1}$.72 109	0.27 891	$\bar{1}$.94 694	$\bar{1}$.68 213	$\bar{1}$.73 927	0.26 073	$\bar{1}$.94 286	15
46	66 827	72 140	27 860	94 687	68 237	73 957	26 043	94 279	14
47	66 851	72 170	27 830	94 680	68 260	73 987	26 013	94 273	13
48	66 875	72 201	27 799	94 674	68 283	74 017	25 983	94 266	12
49	66 899	72 231	27 769	94 667	68 305	74 047	25 953	94 259	11
50	$\bar{1}$.66 922	$\bar{1}$.72 262	0.27 738	$\bar{1}$.94 660	$\bar{1}$.68 328	$\bar{1}$.74 077	0.25 923	$\bar{1}$.94 252	10
51	66 946	72 293	27 707	94 654	68 351	74 107	25 893	94 245	9
52	66 970	72 323	27 677	94 647	68 374	74 137	25 863	94 238	8
53	66 994	72 354	27 646	94 640	68 397	74 166	25 834	94 231	7
54	67 018	72 384	27 616	94 634	68 420	74 196	25 804	94 224	6
55	$\bar{1}$.67 042	$\bar{1}$.72 415	0.27 585	$\bar{1}$.94 627	$\bar{1}$.68 443	$\bar{1}$.74 226	0.25 774	$\bar{1}$.94 217	5
56	67 066	72 445	27 555	94 620	68 466	74 256	25 744	94 210	4
57	67 090	72 476	27 524	94 614	68 489	74 286	25 714	94 203	3
58	67 113	72 506	27 494	94 607	68 512	74 316	25 684	94 196	2
59	67 137	72 537	27 463	94 600	68 534	74 345	25 655	94 189	1
60′	$\bar{1}$.67 161	$\bar{1}$.72 567	0.27 433	$\bar{1}$.94 593	$\bar{1}$.68 557	$\bar{1}$.74 375	0.25 625	$\bar{1}$.94 182	0′
	log cos	log cot	log tan	log sin 62°	log cos	log cot	log tan	log sin 61°	′

	29°				30°				
′	log sin	log tan	log cot	log cos	log sin	log tan	log cot	log cos	
0′	1̄.68 557	1̄.74 375	0.25 625	1̄.94 182	1̄.69 897	1̄.76 144	0.23 856	1̄.93 753	60′
1	68 580	74 405	25 595	94 175	69 919	76 173	23 827	93 746	59
2	68 603	74 435	25 565	94 168	69 941	76 202	23 798	93 738	58
3	68 625	74 465	25 535	94 161	69 963	76 231	23 769	93 731	57
4	68 648	74 494	25 506	94 154	69 984	76 261	23 739	93 724	56
5	1̄.68 671	1̄.74 524	0.25 476	1̄.94 147	1̄.70 006	1̄.76 290	0.23 710	1̄.93 717	55
6	68 694	74 554	25 446	94 140	70 028	76 319	23 681	93 709	54
7	68 716	74 583	25 417	94 133	70 050	76 348	23 652	93 702	53
8	68 739	74 613	25 387	94 126	70 072	76 377	23 623	93 695	52
9	68 762	74 643	25 357	94 119	70 093	76 406	23 594	93 687	51
10	1̄.68 784	1̄.74 673	0.25 327	1̄.94 112	1̄.70 115	1̄.76 435	0.23 565	1̄.93 680	50
11	68 807	74 702	25 298	94 105	70 137	76 464	23 536	93 673	49
12	68 829	74 732	25 268	94 098	70 159	76 493	23 507	93 665	48
13	68 852	74 762	25 238	94 090	70 180	76 522	23 478	93 658	47
14	68 875	74 791	25 209	94 083	70 202	76 551	23 449	93 650	46
15	1̄.68 897	1̄.74 821	0.25 179	1̄.94 076	1̄.70 224	1̄.76 580	0.23 420	1̄.93 643	45
16	68 920	74 851	25 149	94 069	70 245	76 609	23 391	93 636	44
17	68 942	74 880	25 120	94 062	70 267	76 639	23 361	93 628	43
18	68 965	74 910	25 090	94 055	70 288	76 668	23 332	93 621	42
19	68 987	74 939	25 061	94 048	70 310	76 697	23 303	93 614	41
20	1̄.69 010	1̄.74 969	0.25 031	1̄.94 041	1̄.70 332	1̄.76 725	0.23 275	1̄.93 606	40
21	69 032	74 998	25 002	94 034	70 353	76 754	23 246	93 599	39
22	69 055	75 028	24 972	94 027	70 375	76 783	23 217	93 591	38
23	69 077	75 058	24 942	94 020	70 396	76 812	23 188	93 584	37
24	69 100	75 087	24 913	94 012	70 418	76 841	23 159	93 577	36
25	1̄.69 122	1̄.75 117	0.24 883	1̄.94 005	1̄.70 439	1̄.76 870	0.23 130	1̄.93 569	35
26	69 144	75 146	24 854	93 998	70 461	76 899	23 101	93 562	34
27	69 167	75 176	24 824	93 991	70 482	76 928	23 072	93 554	33
28	69 189	75 205	24 795	93 984	70 504	76 957	23 043	93 547	32
29	69 212	75 235	24 765	93 977	70 525	76 986	23 014	93 539	31
30	1̄.69 234	1̄.75 264	0.24 736	1̄.93 970	1̄.70 547	1̄.77 015	0.22 985	1̄.93 532	30
31	69 256	75 294	24 706	93 963	70 568	77 044	22 956	93 525	29
32	69 279	75 323	24 677	93 955	70 590	77 073	22 927	93 517	28
33	69 301	75 353	24 647	93 948	70 611	77 101	22 899	93 510	27
34	69 323	75 382	24 618	93 941	70 633	77 130	22 870	93 502	26
35	1̄.69 345	1̄.75 411	0.24 589	1̄.93 934	1̄.70 654	1̄.77 159	0.22 841	1̄.93 495	25
36	69 368	75 441	24 559	93 927	70 675	77 188	22 812	93 487	24
37	69 390	75 470	24 530	93 920	70 697	77 217	22 783	93 480	23
38	69 412	75 500	24 500	93 912	70 718	77 246	22 754	93 472	22
39	69 434	75 529	24 471	93 905	70 739	77 274	22 726	93 465	21
40	1̄.69 456	1̄.75 558	0.24 442	1̄.93 898	1̄.70 761	1̄.77 303	0.22 697	1̄.93 457	20
41	69 479	75 588	24 412	93 891	70 782	77 332	22 668	93 450	19
42	69 501	75 617	24 383	93 884	70 803	77 361	22 639	93 442	18
43	69 523	75 647	24 353	93 876	70 824	77 390	22 610	93 435	17
44	69 545	75 676	24 324	93 869	70 846	77 418	22 582	93 427	16
45	1̄.69 567	1̄.75 705	0.24 295	1̄.93 862	1̄.70 867	1̄.77 447	0.22 553	1̄.93 420	15
46	69 589	75 735	24 265	93 855	70 888	77 476	22 524	93 412	14
47	69 611	75 764	24 236	93 847	70 909	77 505	22 495	93 405	13
48	69 633	75 793	24 207	93 840	70 931	77 533	22 467	93 397	12
49	69 655	75 822	24 178	93 833	70 952	77 562	22 438	93 390	11
50	1̄.69 677	1̄.75 852	0.24 148	1̄.93 826	1̄.70 973	1̄.77 591	0.22 409	1̄.93 382	10
51	69 699	75 881	24 119	93 819	70 994	77 619	22 381	93 375	9
52	69 721	75 910	24 090	93 811	71 015	77 648	22 352	93 367	8
53	69 743	75 939	24 061	93 804	71 036	77 677	22 323	93 360	7
54	69 765	75 969	24 031	93 797	71 058	77 706	22 294	93 352	6
55	1̄.69 787	1̄.75 998	0.24 002	1̄.93 789	1̄.71 079	1̄.77 734	0.22 266	1̄.93 344	5
56	69 809	76 027	23 973	93 782	71 100	77 763	22 237	93 337	4
57	69 831	76 056	23 944	93 775	71 121	77 791	22 209	93 329	3
58	69 853	76 086	23 914	93 768	71 142	77 820	22 180	93 322	2
59	69 875	76 115	23 885	93 760	71 163	77 849	22 151	93 314	1
60′	1̄.69 897	1̄.76 144	0.23 856	1̄.93 753	1̄.71 184	1̄.77 877	0.22 123	1̄.93 307	0′
	log cos	log cot	log tan	log sin	log cos	log cot	log tan	log sin	′
	60°				59°				

′	31° log sin	log tan	log cot	log cos	32° log sin	log tan	log cot	log cos	
0′	$\bar{1}$.71 184	$\bar{1}$.77 877	0.22 123	$\bar{1}$.93 307	$\bar{1}$.72 421	$\bar{1}$.79 579	0.20 421	$\bar{1}$.92 842	60′
1	71 205	77 906	22 094	93 299	72 441	79 607	20 393	92 834	59
2	71 226	77 935	22 065	93 291	72 461	79 635	20 365	92 826	58
3	71 247	77 963	22 037	93 284	72 482	79 663	20 337	92 818	57
4	71 268	77 992	22 008	93 276	72 502	79 691	20 309	92 810	56
5	$\bar{1}$.71 289	$\bar{1}$.78 020	0.21 980	$\bar{1}$.93 269	$\bar{1}$.72 522	$\bar{1}$.79 719	0.20 281	$\bar{1}$.92 803	55
6	71 310	78 049	21 951	93 261	72 542	79 747	20 253	92 795	54
7	71 331	78 077	21 923	93 253	72 562	79 776	20 224	92 787	53
8	71 352	78 106	21 894	93 246	72 582	79 804	20 196	92 779	52
9	71 373	78 135	21 865	93 238	72 602	79 832	20 168	92 771	51
10	$\bar{1}$.71 393	$\bar{1}$.78 163	0.21 837	$\bar{1}$.93 230	$\bar{1}$.72 622	$\bar{1}$.79 860	0.20 140	$\bar{1}$.92 763	50
11	71 414	78 192	21 808	93 223	72 643	79 888	20 112	92 755	49
12	71 435	78 220	21 780	93 215	72 663	79 916	20 084	92 747	48
13	71 456	78 249	21 751	93 207	72 683	79 944	20 056	92 739	47
14	71 477	78 277	21 723	93 200	72 703	79 972	20 028	92 731	46
15	$\bar{1}$.71 498	$\bar{1}$.78 306	0.21 694	$\bar{1}$.93 192	$\bar{1}$.72 723	$\bar{1}$.80 000	0.20 000	$\bar{1}$.92 723	45
16	71 519	78 334	21 666	93 184	72 743	80 028	19 972	92 715	44
17	71 539	78 363	21 637	93 177	72 763	80 056	19 944	92 707	43
18	71 560	78 391	21 609	93 169	72 783	80 084	19 916	92 699	42
19	71 581	78 419	21 581	93 161	72 803	80 112	19 888	92 691	41
20	$\bar{1}$.71 602	$\bar{1}$.78 448	0.21 552	$\bar{1}$.93 154	$\bar{1}$.72 823	$\bar{1}$.80 140	0.19 860	$\bar{1}$.92 683	40
21	71 622	78 476	21 524	93 146	72 843	80 168	19 832	92 675	39
22	71 643	78 505	21 495	93 138	72 863	80 195	19 805	92 667	38
23	71 664	78 533	21 467	93 131	72 883	80 223	19 777	92 659	37
24	71 685	78 562	21 438	93 123	72 902	80 251	19 749	92 651	36
25	$\bar{1}$.71 705	$\bar{1}$.78 590	0.21 410	$\bar{1}$.93 115	$\bar{1}$.72 922	$\bar{1}$.80 279	0.19 721	$\bar{1}$.92 643	35
26	71 726	78 618	21 382	93 108	72 942	80 307	19 693	92 635	34
27	71 747	78 647	21 353	93 100	72 962	80 335	19 665	92 627	33
28	71 767	78 675	21 325	93 092	72 982	80 363	19 637	92 619	32
29	71 788	78 704	21 296	93 084	73 002	80 391	19 609	92 611	31
30	$\bar{1}$.71 809	$\bar{1}$.78 732	0.21 268	$\bar{1}$.93 077	$\bar{1}$.73 022	$\bar{1}$.80 419	0.19 581	$\bar{1}$.92 603	30
31	71 829	78 760	21 240	93 069	73 041	80 447	19 553	92 595	29
32	71 850	78 789	21 211	93 061	73 061	80 474	19 526	92 587	28
33	71 870	78 817	21 183	93 053	73 081	80 502	19 498	92 579	27
34	71 891	78 845	21 155	93 046	73 101	80 530	19 470	92 571	26
35	$\bar{1}$.71 911	$\bar{1}$.78 874	0.21 126	$\bar{1}$.93 038	$\bar{1}$.73 121	$\bar{1}$.80 558	0.19 442	$\bar{1}$.92 563	25
36	71 932	78 902	21 098	93 030	73 140	80 586	19 414	92 555	24
37	71 952	78 930	21 070	93 022	73 160	80 614	19 386	92 546	23
38	71 973	78 959	21 041	93 014	73 180	80 642	19 358	92 538	22
39	71 994	78 987	21 013	93 007	73 200	80 669	19 331	92 530	21
40	$\bar{1}$.72 014	$\bar{1}$.79 015	0.20 985	$\bar{1}$.92 999	$\bar{1}$.73 219	$\bar{1}$.80 697	0.19 303	$\bar{1}$.92 522	20
41	72 034	79 043	20 957	92 991	73 239	80 725	19 275	92 514	19
42	72 055	79 072	20 928	92 983	73 259	80 753	19 247	92 506	18
43	72 075	79 100	20 900	92 976	73 278	80 781	19 219	92 498	17
44	72 096	79 128	20 872	92 968	73 298	80 808	19 192	92 490	16
45	$\bar{1}$.72 116	$\bar{1}$.79 156	0.20 844	$\bar{1}$.92 960	$\bar{1}$.73 318	$\bar{1}$.80 836	0.19 164	$\bar{1}$.92 482	15
46	72 137	79 185	20 815	92 952	73 337	80 864	19 136	92 473	14
47	72 157	79 213	20 787	92 944	73 357	80 892	19 108	92 465	13
48	72 177	79 241	20 759	92 936	73 377	80 919	19 081	92 457	12
49	72 198	79 269	20 731	92 929	73 396	80 947	19 053	92 449	11
50	$\bar{1}$.72 218	$\bar{1}$.79 297	0.20 703	$\bar{1}$.92 921	$\bar{1}$.73 416	$\bar{1}$.80 975	0.19 025	$\bar{1}$.92 441	10
51	72 238	79 326	20 674	92 913	73 435	81 003	18 997	92 433	9
52	72 259	79 354	20 646	92 905	73 455	81 030	18 970	92 425	8
53	72 279	79 382	20 618	92 897	73 474	81 058	18 942	92 416	7
54	72 299	79 410	20 590	92 889	73 494	81 086	18 914	92 408	6
55	$\bar{1}$.72 320	$\bar{1}$.79 438	0.20 562	$\bar{1}$.92 881	$\bar{1}$.73 513	$\bar{1}$.81 113	0.18 887	$\bar{1}$.92 400	5
56	72 340	79 466	20 534	92 874	73 533	81 141	18 859	92 392	4
57	72 360	79 495	20 505	92 866	73 552	81 169	18 831	92 384	3
58	72 381	79 523	20 477	92 858	73 572	81 196	18 804	92 376	2
59	72 401	79 551	20 449	92 850	73 591	81 224	18 776	92 367	1
60′	$\bar{1}$.72 421	$\bar{1}$.79 579	0.20 421	$\bar{1}$.92 842	$\bar{1}$.73 611	$\bar{1}$.81 252	0.18 748	$\bar{1}$.92 359	0′
	log cos	log cot	log tan	log sin	log cos	log cot	log tan	log sin	′
	58°				57°				

33° | 34°

′	log sin	log tan	log cot	log cos	log sin	log tan	log cot	log cos	
0′	1̄.73 611	1̄.81 252	0.18 748	1̄.92 359	1̄.74 756	1̄.82 899	0.17 101	1̄.91 857	60′
1	73 630	81 279	18 721	92 351	74 775	82 926	17 074	91 849	59
2	73 650	81 307	18 693	92 343	74 794	82 953	17 047	91 840	58
3	73 669	81 335	18 665	92 335	74 812	82 980	17 020	91 832	57
4	73 689	81 362	18 638	92 326	74 831	83 008	16 992	91 823	56
5	1̄.73 708	1̄.81 390	0.18 610	1̄.92 318	1̄.74 850	1̄.83 035	0.16 965	1̄.91 815	55
6	73 727	81 418	18 582	92 310	74 868	83 062	16 938	91 806	54
7	73 747	81 445	18 555	92 302	74 887	83 089	16 911	91 798	53
8	73 766	81 473	18 527	92 293	74 906	83 117	16 883	91 789	52
9	73 785	81 500	18 500	92 285	74 924	83 144	16 856	91 781	51
10	1̄.73 805	1̄.81 528	0.18 472	1̄.92 277	1̄.74 943	1̄.83 171	0.16 829	1̄.91 772	50
11	73 824	81 556	18 444	92 269	74 961	83 198	16 802	91 763	49
12	73 843	81 583	18 417	92 260	74 980	83 225	16 775	91 755	48
13	73 863	81 611	18 389	92 252	74 999	83 252	16 748	91 746	47
14	73 882	81 638	18 362	92 244	75 017	83 280	16 720	91 738	46
15	1̄.73 901	1̄.81 666	0.18 334	1̄.92 235	1̄.75 036	1̄.83 307	0.16 693	1̄.91 729	45
16	73 921	81 693	18 307	92 227	75 054	83 334	16 666	91 720	44
17	73 940	81 721	18 279	92 219	75 073	83 361	16 639	91 712	43
18	73 959	81 748	18 252	92 211	75 091	83 388	16 612	91 703	42
19	73 978	81 776	18 224	92 202	75 110	83 415	16 585	91 695	41
20	1̄.73 997	1̄.81 803	0.18 197	1̄.92 194	1̄.75 128	1̄.83 442	0.16 558	1̄.91 686	40
21	74 017	81 831	18 169	92 186	75 147	83 470	16 530	91 677	39
22	74 036	81 858	18 142	92 177	75 165	83 497	16 503	91 669	38
23	74 055	81 886	18 114	92 169	75 184	83 524	16 476	91 660	37
24	74 074	81 913	18 087	92 161	75 202	83 551	16 449	91 651	36
25	1̄.74 093	1̄.81 941	0.18 059	1̄.92 152	1̄.75 221	1̄.83 578	0.16 422	1̄.91 643	35
26	74 113	81 968	18 032	92 144	75 239	83 605	16 395	91 634	34
27	74 132	81 996	18 004	92 136	75 258	83 632	16 368	91 625	33
28	74 151	82 023	17 977	92 127	75 276	83 659	16 341	91 617	32
29	74 170	82 051	17 949	92 119	75 294	83 686	16 314	91 608	31
30	1̄.74 189	1̄.82 078	0.17 922	1̄.92 111	1̄.75 313	1̄.83 713	0.16 287	1̄.91 599	30
31	74 208	82 106	17 894	92 102	75 331	83 740	16 260	91 591	29
32	74 227	82 133	17 867	92 094	75 350	83 768	16 232	91 582	28
33	74 246	82 161	17 839	92 086	75 368	83 795	16 205	91 573	27
34	74 265	82 188	17 812	92 077	75 386	83 822	16 178	91 565	26
35	1̄.74 284	1̄.82 215	0.17 785	1̄.92 069	1̄.75 405	1̄.83 849	0.16 151	1̄.91 556	25
36	74 303	82 243	17 757	92 060	75 423	83 876	16 124	91 547	24
37	74 322	82 270	17 730	92 052	75 441	83 903	16 097	91 538	23
38	74 341	82 298	17 702	92 044	75 459	83 930	16 070	91 530	22
39	74 360	82 325	17 675	92 035	75 478	83 957	16 043	91 521	21
40	1̄.74 379	1̄.82 352	0.17 648	1̄.92 027	1̄.75 496	1̄.83 984	0.16 016	1̄.91 512	20
41	74 398	82 380	17 620	92 018	75 514	84 011	15 989	91 504	19
42	74 417	82 407	17 593	92 010	75 533	84 038	15 962	91 495	18
43	74 436	82 435	17 565	92 002	75 551	84 065	15 935	91 486	17
44	74 455	82 462	17 538	91 993	75 569	84 092	15 908	91 477	16
45	1̄.74 474	1̄.82 489	0.17 511	1̄.91 985	1̄.75 587	1̄.84 119	0.15 881	1̄.91 469	15
46	74 493	82 517	17 483	91 976	75 605	84 146	15 854	91 460	14
47	74 512	82 544	17 456	91 968	75 624	84 173	15 827	91 451	13
48	74 531	82 571	17 429	91 959	75 642	84 200	15 800	91 442	12
49	74 549	82 599	17 401	91 951	75 660	84 227	15 773	91 433	11
50	1̄.74 568	1̄.82 626	0.17 374	1̄.91 942	1̄.75 678	1̄.84 254	0.15 746	1̄.91 425	10
51	74 587	82 653	17 347	91 934	75 696	84 280	15 720	91 416	9
52	74 606	82 681	17 319	91 925	75 714	84 307	15 693	91 407	8
53	74 625	82 708	17 292	91 917	75 733	84 334	15 666	91 398	7
54	74 644	82 735	17 265	91 908	75 751	84 361	15 639	91 389	6
55	1̄.74 662	1̄.82 762	0.17 238	1̄.91 900	1̄.75 769	1̄.84 388	0.15 612	1̄.91 381	5
56	74 681	82 790	17 210	91 891	75 787	84 415	15 585	91 372	4
57	74 700	82 817	17 183	91 883	75 805	84 442	15 558	91 363	3
58	74 719	82 844	17 156	91 874	75 823	84 469	15 531	91 354	2
59	74 737	82 871	17 129	91 866	75 841	84 496	15 504	91 345	1
60′	1̄.74 756	1̄.82 899	0.17 101	1̄.91 857	1̄.75 859	1̄.84 523	0.15 477	1̄.91 336	0′
	log cos	log cot	log tan	log sin	log cos	log cot	log tan	log sin	′

56° | 55°

	35°				36°				
′	log sin	log tan	log cot	log cos	log sin	log tan	log cot	log cos	
0′	$\bar{1}$.75 859	$\bar{1}$.84 523	0.15 477	$\bar{1}$.91 336	$\bar{1}$.76 922	$\bar{1}$.86 126	0.13 874	$\bar{1}$.90 796	60′
1	75 877	84 550	15 450	91 328	76 939	86 153	13 847	90 787	59
2	75 895	84 576	15 424	91 319	76 957	86 179	13 821	90 777	58
3	75 913	84 603	15 397	91 310	76 974	86 206	13 794	90 768	57
4	75 931	84 630	15 370	91 301	76 991	86 232	13 768	90 759	56
5	$\bar{1}$.75 949	$\bar{1}$.84 657	0.15 343	$\bar{1}$.91 292	$\bar{1}$.77 009	$\bar{1}$.86 259	0.13 741	$\bar{1}$.90 750	55
6	75 967	84 684	15 316	91 283	77 026	86 285	13 715	90 741	54
7	75 985	84 711	15 289	91 274	77 043	86 312	13 688	90 731	53
8	76 003	84 738	15 262	91 266	77 061	86 338	13 662	90 722	52
9	76 021	84 764	15 236	91 257	77 078	86 365	13 635	90 713	51
10	$\bar{1}$.76 039	$\bar{1}$.84 791	0.15 209	$\bar{1}$.91 248	$\bar{1}$.77 095	$\bar{1}$.86 392	0.13 608	$\bar{1}$.90 704	50
11	76 057	84 818	15 182	91 239	77 112	86 418	13 582	90 694	49
12	76 075	84 845	15 155	91 230	77 130	86 445	13 555	90 685	48
13	76 093	84 872	15 128	91 221	77 147	86 471	13 529	90 676	47
14	76 111	84 899	15 101	91 212	77 164	86 498	13 502	90 667	46
15	$\bar{1}$.76 129	$\bar{1}$.84 925	0.15 075	$\bar{1}$.91 203	$\bar{1}$.77 181	$\bar{1}$.86 524	0.13 476	$\bar{1}$.90 657	45
16	76 146	84 952	15 048	91 194	77 199	86 551	13 449	90 648	44
17	76 164	84 979	15 021	91 185	77 216	86 577	13 423	90 639	43
18	76 182	85 006	14 994	91 176	77 233	86 603	13 397	90 630	42
19	76 200	85 033	14 967	91 167	77 250	86 630	13 370	90 620	41
20	$\bar{1}$.76 218	$\bar{1}$.85 059	0.14 941	$\bar{1}$.91 158	$\bar{1}$.77 268	$\bar{1}$.86 656	0.13 344	$\bar{1}$.90 611	40
21	76 236	85 086	14 914	91 149	77 285	86 683	13 317	90 602	39
22	76 253	85 113	14 887	91 141	77 302	86 709	13 291	90 592	38
23	76 271	85 140	14 860	91 132	77 319	86 736	13 264	90 583	37
24	76 289	85 166	14 834	91 123	77 336	86 762	13 238	90 574	36
25	$\bar{1}$.76 307	$\bar{1}$.85 193	0.14 807	$\bar{1}$.91 114	$\bar{1}$.77 353	$\bar{1}$.86 789	0.13 211	$\bar{1}$.90 565	35
26	76 324	85 220	14 780	91 105	77 370	86 815	13 185	90 555	34
27	76 342	85 247	14 753	91 096	77 387	86 842	13 158	90 546	33
28	76 360	85 273	14 727	91 087	77 405	86 868	13 132	90 537	32
29	76 378	85 300	14 700	91 078	77 422	86 894	13 106	90 527	31
30	$\bar{1}$.76 395	$\bar{1}$.85 327	0.14 673	$\bar{1}$.91 069	$\bar{1}$.77 439	$\bar{1}$.86 921	0.13 079	$\bar{1}$.90 518	30
31	76 413	85 354	14 646	91 060	77 456	86 947	13 053	90 509	29
32	76 431	85 380	14 620	91 051	77 473	86 974	13 026	90 499	28
33	76 448	85 407	14 593	91 042	77 490	87 000	13 000	90 490	27
34	76 466	85 434	14 566	91 033	77 507	87 027	12 973	90 480	26
35	$\bar{1}$.76 484	$\bar{1}$.85 460	0.14 540	$\bar{1}$.91 023	$\bar{1}$.77 524	$\bar{1}$.87 053	0.12 947	$\bar{1}$.90 471	25
36	76 501	85 487	14 513	91 014	77 541	87 079	12 921	90 462	24
37	76 519	85 514	14 486	91 005	77 558	87 106	12 894	90 452	23
38	76 537	85 540	14 460	90 996	77 575	87 132	12 868	90 443	22
39	76 554	85 567	14 433	90 987	77 592	87 158	12 842	90 434	21
40	$\bar{1}$.76 572	$\bar{1}$.85 594	0.14 406	$\bar{1}$.90 978	$\bar{1}$.77 609	$\bar{1}$.87 185	0.12 815	$\bar{1}$.90 424	20
41	76 590	85 620	14 380	90 969	77 626	87 211	12 789	90 415	19
42	76 607	85 647	14 353	90 960	77 643	87 238	12 762	90 405	18
43	76 625	85 674	14 326	90 951	77 660	87 264	12 736	90 396	17
44	76 642	85 700	14 300	90 942	77 677	87 290	12 710	90 386	16
45	$\bar{1}$.76 660	$\bar{1}$.85 727	0.14 273	$\bar{1}$.90 933	$\bar{1}$.77 694	$\bar{1}$.87 317	0.12 683	$\bar{1}$.90 377	15
46	76 677	85 754	14 246	90 924	77 711	87 343	12 657	90 368	14
47	76 695	85 780	14 220	90 915	77 728	87 369	12 631	90 358	13
48	76 712	85 807	14 193	90 906	77 744	87 396	12 604	90 349	12
49	76 730	85 834	14 166	90 896	77 761	87 422	12 578	90 339	11
50	$\bar{1}$.76 747	$\bar{1}$.85 860	0.14 140	$\bar{1}$.90 887	$\bar{1}$.77 778	$\bar{1}$.87 448	0.12 552	$\bar{1}$.90 330	10
51	76 765	85 887	14 113	90 878	77 795	87 475	12 525	90 320	9
52	76 782	85 913	14 087	90 869	77 812	87 501	12 499	90 311	8
53	76 800	85 940	14 060	90 860	77 829	87 527	12 473	90 301	7
54	76 817	85 967	14 033	90 851	77 846	87 554	12 446	90 292	6
55	$\bar{1}$.76 835	$\bar{1}$.85 993	0.14 007	$\bar{1}$.90 842	$\bar{1}$.77 862	$\bar{1}$.87 580	0.12 420	$\bar{1}$.90 282	5
56	76 852	86 020	13 980	90 832	77 879	87 606	12 394	90 273	4
57	76 870	86 046	13 954	90 823	77 896	87 633	12 367	90 263	3
58	76 887	86 073	13 927	90 814	77 913	87 659	12 341	90 254	2
59	76 904	86 100	13 900	90 805	77 930	87 685	12 315	90 244	1
60′	$\bar{1}$.76 922	$\bar{1}$.86 126	0.13 874	$\bar{1}$.90 796	$\bar{1}$.77 946	$\bar{1}$.87 711	0.12 289	$\bar{1}$.90 235	0′
	log cos	log cot	log tan	log sin	log cos	log cot	log tan	log sin	′
	54°				53°				

	37°				38°				
′	log sin	log tan	log cot	log cos	log sin	log tan	log cot	log cos	
0′	1̄.77 946	1̄.87 711	0.12 289	1̄.90 235	1̄.78 934	1̄.89 281	0.10 719	1̄.89 653	60
1	77 963	87 738	12 262	90 225	78 950	89 307	10 693	89 643	59
2	77 980	87 764	12 236	90 216	78 967	89 333	10 667	89 633	58
3	77 997	87 790	12 210	90 206	78 983	89 359	10 641	89 624	57
4	78 013	87 817	12 183	90 197	78 999	89 385	10 615	89 614	56
5	1̄.78 030	1̄.87 843	0.12 157	1̄.90 187	1̄.79 015	1̄.89 411	0.10 589	1̄.89 604	55
6	78 047	87 869	12 131	90 178	79 031	89 437	10 563	89 594	54
7	78 063	87 895	12 105	90 168	79 047	89 463	10 537	89 584	53
8	78 080	87 922	12 078	90 159	79 063	89 489	10 511	89 574	52
9	78 097	87 948	12 052	90 149	79 079	89 515	10 485	89 564	51
10	1̄.78 113	1̄.87 974	0.12 026	1̄.90 139	1̄.79 095	1̄.89 541	0.10 459	1̄.89 554	50
11	78 130	88 000	12 000	90 130	79 111	89 567	10 433	89 544	49
12	78 147	88 027	11 973	90 120	79 128	89 593	10 407	89 534	48
13	78 163	88 053	11 947	90 111	79 144	89 619	10 381	89 524	47
14	78 180	88 079	11 921	90 101	79 160	89 645	10 355	89 514	46
15	1̄.78 197	1̄.88 105	0.11 895	1̄.90 091	1̄.79 176	1̄.89 671	0.10 329	1̄.89 504	45
16	78 213	88 131	11 869	90 082	79 192	89 697	10 303	89 495	44
17	78 230	88 158	11 842	90 072	79 208	89 723	10 277	89 485	43
18	78 246	88 184	11 816	90 063	79 224	89 749	10 251	89 475	42
19	78 263	88 210	11 790	90 053	79 240	89 775	10 225	89 465	41
20	1̄.78 280	1̄.88 236	0.11 764	1̄.90 043	1̄.79 256	1̄.89 801	0.10 199	1̄.89 455	40
21	78 296	88 262	11 738	90 034	79 272	89 827	10 173	89 445	39
22	78 313	88 289	11 711	90 024	79 288	89 853	10 147	89 435	38
23	78 329	88 315	11 685	90 014	79 304	89 879	10 121	89 425	37
24	78 346	88 341	11 659	90 005	79 319	89 905	10 095	89 415	36
25	1̄.78 362	1̄.88 367	0.11 633	1̄.89 995	1̄.79 335	1̄.89 931	0.10 069	1̄.89 405	35
26	78 379	88 393	11 607	89 985	79 351	89 957	10 043	89 395	34
27	78 395	88 420	11 580	89 976	79 367	89 983	10 017	89 385	33
28	78 412	88 446	11 554	89 966	79 383	90 009	09 991	89 375	32
29	78 428	88 472	11 528	89 956	79 399	90 035	09 965	89 364	31
30	1̄.78 445	1̄.88 498	0.11 502	1̄.89 947	1̄.79 415	1̄.90 061	0.09 939	1̄.89 354	30
31	78 461	88 524	11 476	89 937	79 431	90 086	09 914	89 344	29
32	78 478	88 550	11 450	89 927	79 447	90 112	09 888	89 334	28
33	78 494	88 577	11 423	89 918	79 463	90 138	09 862	89 324	27
34	78 510	88 603	11 397	89 908	79 478	90 164	09 836	89 314	26
35	1̄.78 527	1̄.88 629	0.11 371	1̄.89 898	1̄.79 494	1̄.90 190	0.09 810	1̄.89 304	25
36	78 543	88 655	11 345	89 888	79 510	90 216	09 784	89 294	24
37	78 560	88 681	11 319	89 879	79 526	90 242	09 758	89 284	23
38	78 576	88 707	11 293	89 869	79 542	90 268	09 732	89 274	22
39	78 592	88 733	11 267	89 859	79 558	90 294	09 706	89 264	21
40	1̄.78 609	1̄.88 759	0.11 241	1̄.89 849	1̄.79 573	1̄.90 320	0.09 680	1̄.89 254	20
41	78 625	88 786	11 214	89 840	79 589	90 346	09 654	89 244	19
42	78 642	88 812	11 188	89 830	79 605	90 371	09 629	89 233	18
43	78 658	88 838	11 162	89 820	79 621	90 397	09 603	89 223	17
44	78 674	88 864	11 136	89 810	79 636	90 423	09 577	89 213	16
45	1̄.78 691	1̄.88 890	0.11 110	1̄.89 801	1̄.79 652	1̄.90 449	0.09 551	1̄.89 203	15
46	78 707	88 916	11 084	89 791	79 668	90 475	09 525	89 193	14
47	78 723	88 942	11 058	89 781	79 684	90 501	09 499	89 183	13
48	78 739	88 968	11 032	89 771	79 699	90 527	09 473	89 173	12
49	78 756	88 994	11 006	89 761	79 715	90 553	09 447	89 162	11
50	1̄.78 772	1̄.89 020	0.10 980	1̄.89 752	1̄.79 731	1̄.90 578	0.09 422	1̄.89 152	10
51	78 788	89 046	10 954	89 742	79 746	90 604	09 396	89 142	9
52	78 805	89 073	10 927	89 732	79 762	90 630	09 370	89 132	8
53	78 821	89 099	10 901	89 722	79 778	90 656	09 344	89 122	7
54	78 837	89 125	10 875	89 712	79 793	90 682	09 318	89 112	6
55	1̄.78 853	1̄.89 151	0.10 849	1̄.89 702	1̄.79 809	1̄.90 708	0.09 292	1̄.89 101	5
56	78 869	89 177	10 823	89 693	79 825	90 734	09 266	89 091	4
57	78 886	89 203	10 797	89 683	79 840	90 759	09 241	89 081	3
58	78 902	89 229	10 771	89 673	79 856	90 785	09 215	89 071	2
59	78 918	89 255	10 745	89 663	79 872	90 811	09 189	89 060	1
60′	1̄.78 934	1̄.89 281	0.10 719	1̄.89 653	1̄.79 887	1̄.90 837	0.09 163	1̄.89 050	0′
	log cos	log cot	log tan	log sin	log cos	log cot	log tan	log sin	′
	52°				51°				

′	39° log sin	log tan	log cot	log cos	40° log sin	log tan	log cot	log cos	′
0′	$\bar{1}$.79 887	$\bar{1}$.90 837	0.09 163	$\bar{1}$.89 050	$\bar{1}$.80 807	$\bar{1}$.92 381	0.07 619	$\bar{1}$.88 425	60′
1	79 903	90 863	09 137	89 040	80 822	92 407	07 593	88 415	59
2	79 918	90 889	09 111	89 030	80 837	92 433	07 567	88 404	58
3	79 934	90 914	09 086	89 020	80 852	92 458	07 542	88 394	57
4	79 950	90 940	09 060	89 009	80 867	92 484	07 516	88 383	56
5	$\bar{1}$.79 965	$\bar{1}$.90 966	0.09 034	$\bar{1}$.88 999	$\bar{1}$.80 882	$\bar{1}$.92 510	0.07 490	$\bar{1}$.88 372	55
6	79 981	90 992	09 008	88 989	80 897	92 535	07 465	88 362	54
7	79 996	91 018	08 982	88 978	80 912	92 561	07 439	88 351	53
8	80 012	91 043	08 957	88 968	80 927	92 587	07 413	88 340	52
9	80 027	91 069	08 931	88 958	80 942	92 612	07 388	88 330	51
10	$\bar{1}$.80 043	$\bar{1}$.91 095	0.08 905	$\bar{1}$.88 948	$\bar{1}$.80 957	$\bar{1}$.92 638	0.07 362	$\bar{1}$.88 319	50
11	80 058	91 121	08 879	88 937	80 972	92 663	07 337	88 308	49
12	80 074	91 147	08 853	88 927	80 987	92 689	07 311	88 298	48
13	80 089	91 172	08 828	88 917	81 002	92 715	07 285	88 287	47
14	80 105	91 198	08 802	88 906	81 017	92 740	07 260	88 276	46
15	$\bar{1}$.80 120	$\bar{1}$.91 224	0.08 776	$\bar{1}$.88 896	$\bar{1}$.81 032	$\bar{1}$.92 766	0.07 234	$\bar{1}$.88 266	45
16	80 136	91 250	08 750	88 886	81 047	92 792	07 208	88 255	44
17	80 151	91 276	08 724	88 875	81 061	92 817	07 183	88 244	43
18	80 166	91 301	08 699	88 865	81 076	92 843	07 157	88 234	42
19	80 182	91 327	08 673	88 855	81 091	92 868	07 132	88 223	41
20	$\bar{1}$.80 197	$\bar{1}$.91 353	0.08 647	$\bar{1}$.88 844	$\bar{1}$.81 106	$\bar{1}$.92 894	0.07 106	$\bar{1}$.88 212	40
21	80 213	91 379	08 621	88 834	81 121	92 920	07 080	88 201	39
22	80 228	91 404	08 596	88 824	81 136	92 945	07 055	88 191	38
23	80 244	91 430	08 570	88 813	81 151	92 971	07 029	88 180	37
24	80 259	91 456	08 544	88 803	81 166	92 996	07 004	88 169	36
25	$\bar{1}$.80 274	$\bar{1}$.91 482	0.08 518	$\bar{1}$.88 793	$\bar{1}$.81 180	$\bar{1}$.93 022	0.06 978	$\bar{1}$.88 158	35
26	80 290	91 507	08 493	88 782	81 195	93 048	06 952	88 148	34
27	80 305	91 533	08 467	88 772	81 210	93 073	06 927	88 137	33
28	80 320	91 559	08 441	88 761	81 225	93 099	06 901	88 126	32
29	80 336	91 585	08 415	88 751	81 240	93 124	06 876	88 115	31
30	$\bar{1}$.80 351	$\bar{1}$.91 610	0.08 390	$\bar{1}$.88 741	$\bar{1}$.81 254	$\bar{1}$.93 150	0.06 850	$\bar{1}$.88 105	30
31	80 366	91 636	08 364	88 730	81 269	93 175	06 825	88 094	29
32	80 382	91 662	08 338	88 720	81 284	93 201	06 799	88 083	28
33	80 397	91 688	08 312	88 709	81 299	93 227	06 773	88 072	27
34	80 412	91 713	08 287	88 699	81 314	93 252	06 748	88 061	26
35	$\bar{1}$.80 428	$\bar{1}$.91 739	0.08 261	$\bar{1}$.88 688	$\bar{1}$.81 328	$\bar{1}$.93 278	0.06 722	$\bar{1}$.88 051	25
36	80 443	91 765	08 235	88 678	81 343	93 303	06 697	88 040	24
37	80 458	91 791	08 209	88 668	81 358	93 329	06 671	88 029	23
38	80 473	91 816	08 184	88 657	81 372	93 354	06 646	88 018	22
39	80 489	91 842	08 158	88 647	81 387	93 380	06 620	88 007	21
40	$\bar{1}$.80 504	$\bar{1}$.91 868	0.08 132	$\bar{1}$.88 636	$\bar{1}$.81 402	$\bar{1}$.93 406	0.06 594	$\bar{1}$.87 996	20
41	80 519	91 893	08 107	88 626	81 417	93 431	06 569	87 985	19
42	80 534	91 919	08 081	88 615	81 431	93 457	06 543	87 975	18
43	80 550	91 945	08 055	88 605	81 446	93 482	06 518	87 964	17
44	80 565	91 971	08 029	88 594	81 461	93 508	06 492	87 953	16
45	$\bar{1}$.80 580	$\bar{1}$.91 996	0.08 004	$\bar{1}$.88 584	$\bar{1}$.81 475	$\bar{1}$.93 533	0.06 467	$\bar{1}$.87 942	15
46	80 595	92 022	07 978	88 573	81 490	93 559	06 441	87 931	14
47	80 610	92 048	07 952	88 563	81 505	93 584	06 416	87 920	13
48	80 625	92 073	07 927	88 552	81 519	93 610	06 390	87 909	12
49	80 641	92 099	07 901	88 542	81 534	93 636	06 364	87 898	11
50	$\bar{1}$.80 656	$\bar{1}$.92 125	0.07 875	$\bar{1}$.88 531	$\bar{1}$.81 549	$\bar{1}$.93 661	0.06 339	$\bar{1}$.87 887	10
51	80 671	92 150	07 850	88 521	81 563	93 687	06 313	87 877	9
52	80 686	92 176	07 824	88 510	81 578	93 712	06 288	87 866	8
53	80 701	92 202	07 798	88 499	81 592	93 738	06 262	87 855	7
54	80 716	92 227	07 773	88 489	81 607	93 763	06 237	87 844	6
55	$\bar{1}$.80 731	$\bar{1}$.92 253	0.07 747	$\bar{1}$.88 478	$\bar{1}$.81 622	$\bar{1}$.93 789	0.06 211	$\bar{1}$.87 833	5
56	80 746	92 279	07 721	88 468	81 636	93 814	06 186	87 822	4
57	80 762	92 304	07 696	88 457	81 651	93 840	06 160	87 811	3
58	80 777	92 330	07 670	88 447	81 665	93 865	06 135	87 800	2
59	80 792	92 356	07 644	88 436	81 680	93 891	06 109	87 789	1
60′	$\bar{1}$.80 807	$\bar{1}$.92 381	0.07 619	$\bar{1}$.88 425	$\bar{1}$.81 694	$\bar{1}$.93 916	0.06 084	$\bar{1}$.87 778	0′
	50° log cos	log cot	log tan	log sin	49° log cos	log cot	log tan	log sin	′

′	41° log sin	log tan	log cot	log cos	42° log sin	log tan	log cot	log cos	′
0′	$\bar{1}$.81 694	$\bar{1}$.93 916	0.06 084	$\bar{1}$.87 778	$\bar{1}$.82 551	$\bar{1}$.95 444	0.04 556	$\bar{1}$.87 107	60′
1	81 709	93 942	06 058	87 767	82 565	95 469	04 531	87 096	59
2	81 723	93 967	06 033	87 756	82 579	95 495	04 505	87 085	58
3	81 738	93 993	06 007	87 745	82 593	95 520	04 480	87 073	57
4	81 752	94 018	05 982	87 734	82 607	95 545	04 455	87 062	56
5	$\bar{1}$.81 767	$\bar{1}$.94 044	0.05 956	$\bar{1}$.87 723	$\bar{1}$.82 621	$\bar{1}$.95 571	0.04 429	$\bar{1}$.87 050	55
6	81 781	94 069	05 931	87 712	82 635	95 596	04 404	87 039	54
7	81 796	94 095	05 905	87 701	82 649	95 622	04 378	87 028	53
8	81 810	94 120	05 880	87 690	82 663	95 647	04 353	87 016	52
9	81 825	94 146	05 854	87 679	82 677	95 672	04 328	87 005	51
10	$\bar{1}$.81 839	$\bar{1}$.94 171	0.05 829	$\bar{1}$.87 668	$\bar{1}$.82 691	$\bar{1}$.95 698	0.04 302	$\bar{1}$.86 993	50
11	81 854	94 197	05 803	87 657	82 705	95 723	04 277	86 982	49
12	81 868	94 222	05 778	87 646	82 719	95 748	04 252	86 970	48
13	81 882	94 248	05 752	87 635	82 733	95 774	04 226	86 959	47
14	81 897	94 273	05 727	87 624	82 747	95 799	04 201	86 947	46
15	$\bar{1}$.81 911	$\bar{1}$.94 299	0.05 701	$\bar{1}$.87 613	$\bar{1}$.82 761	$\bar{1}$.95 825	0.04 175	$\bar{1}$.86 936	45
16	81 926	94 324	05 676	87 601	82 775	95 850	04 150	86 924	44
17	81 940	94 350	05 650	87 590	82 788	95 875	04 125	86 913	43
18	81 955	94 375	05 625	87 579	82 802	95 901	04 099	86 902	42
19	81 969	94 401	05 599	87 568	82 816	95 926	04 074	86 890	41
20	$\bar{1}$.81 983	$\bar{1}$.94 426	0.05 574	$\bar{1}$.87 557	$\bar{1}$.82 830	$\bar{1}$.95 952	0.04 048	$\bar{1}$.86 879	40
21	81 998	94 452	05 548	87 546	82 844	95 977	04 023	86 867	39
22	82 012	94 477	05 523	87 535	82 858	96 002	03 998	86 855	38
23	82 026	94 503	05 497	87 524	82 872	96 028	03 972	86 844	37
24	82 041	94 528	05 472	87 513	82 885	96 053	03 947	86 832	36
25	$\bar{1}$.82 055	$\bar{1}$.94 554	0.05 446	$\bar{1}$.87 501	$\bar{1}$.82 899	$\bar{1}$.96 078	0.03 922	$\bar{1}$.86 821	35
26	82 069	94 579	05 421	87 490	82 913	96 104	03 896	86 809	34
27	82 084	94 604	05 396	87 479	82 927	96 129	03 871	86 798	33
28	82 098	94 630	05 370	87 468	82 941	96 155	03 845	86 786	32
29	82 112	94 655	05 345	87 457	82 955	96 180	03 820	86 775	31
30	$\bar{1}$.82 126	$\bar{1}$.94 681	0.05 319	$\bar{1}$.87 446	$\bar{1}$.82 968	$\bar{1}$.96 205	0.03 795	$\bar{1}$.86 763	30
31	82 141	94 706	05 294	87 434	82 982	96 231	03 769	86 752	29
32	82 155	94 732	05 268	87 423	82 996	96 256	03 744	86 740	28
33	82 169	94 757	05 243	87 412	83 010	96 281	03 719	86 728	27
34	82 184	94 783	05 217	87 401	83 023	96 307	03 693	86 717	26
35	$\bar{1}$.82 198	$\bar{1}$.94 808	0.05 192	$\bar{1}$.87 390	$\bar{1}$.83 037	$\bar{1}$.96 332	0.03 668	$\bar{1}$.86 705	25
36	82 212	94 834	05 166	87 378	83 051	96 357	03 643	86 694	24
37	82 226	94 859	05 141	87 367	83 065	96 383	03 617	86 682	23
38	82 240	94 884	05 116	87 356	83 078	96 408	03 592	86 670	22
39	82 255	94 910	05 090	87 345	83 092	96 433	03 567	86 659	21
40	$\bar{1}$.82 269	$\bar{1}$.94 935	0.05 065	$\bar{1}$.87 334	$\bar{1}$.83 106	$\bar{1}$.96 459	0.03 541	$\bar{1}$.86 647	20
41	82 283	94 961	05 039	87 322	83 120	96 484	03 516	86 635	19
42	82 297	94 986	05 014	87 311	83 133	96 510	03 490	86 624	18
43	82 311	95 012	04 988	87 300	83 147	96 535	03 465	86 612	17
44	82 326	95 037	04 963	87 288	83 161	96 560	03 440	86 600	16
45	$\bar{1}$.82 340	$\bar{1}$.95 062	0.04 938	$\bar{1}$.87 277	$\bar{1}$.83 174	$\bar{1}$.96 586	0.03 414	$\bar{1}$.86 589	15
46	82 354	95 088	04 912	87 266	83 188	96 611	03 389	86 577	14
47	82 368	95 113	04 887	87 255	83 202	96 636	03 364	86 565	13
48	82 382	95 139	04 861	87 243	83 215	96 662	03 338	86 554	12
49	82 396	95 164	04 836	87 232	83 229	96 687	03 313	86 542	11
50	$\bar{1}$.82 410	$\bar{1}$.95 190	0.04 810	$\bar{1}$.87 221	$\bar{1}$.83 242	$\bar{1}$.96 712	0.03 288	$\bar{1}$.86 530	10
51	82 424	95 215	04 785	87 209	83 256	96 738	03 262	86 518	9
52	82 439	95 240	04 760	87 198	83 270	96 763	03 237	86 507	8
53	82 453	95 266	04 734	87 187	83 283	96 788	03 212	86 495	7
54	82 467	95 291	04 709	87 175	83 297	96 814	03 186	86 483	6
55	$\bar{1}$.82 481	$\bar{1}$.95 317	0.04 683	$\bar{1}$.87 164	$\bar{1}$.83 310	$\bar{1}$.96 839	0.03 161	$\bar{1}$.86 472	5
56	82 495	95 342	04 658	87 153	83 324	96 864	03 136	86 460	4
57	82 509	95 368	04 632	87 141	83 338	96 890	03 110	86 448	3
58	82 523	95 393	04 607	87 130	83 351	96 915	03 085	86 436	2
59	82 537	95 418	04 582	87 119	83 365	96 940	03 060	86 425	1
60′	$\bar{1}$.82 551	$\bar{1}$.95 444	0.04 556	$\bar{1}$.87 107	$\bar{1}$.83 378	$\bar{1}$.96 966	0.03 034	$\bar{1}$.86 413	0′
	log cos	log cot	log tan	log sin 48°	log cos	log cot	log tan	log sin 47°	′

′	43° log sin	log tan	log cot	log cos	44° log sin	log tan	log cot	log cos	′
0′	$\bar{1}$.83 378	$\bar{1}$.96 966	0.03 034	$\bar{1}$.86 413	$\bar{1}$.84 177	$\bar{1}$.98 484	0.01 516	$\bar{1}$.85 693	60′
1	83 392	96 991	03 009	86 401	84 190	98 509	01 491	85 681	59
2	83 405	97 016	02 984	86 389	84 203	98 534	01 466	85 669	58
3	83 419	97 042	02 958	86 377	84 216	98 560	01 440	85 657	57
4	83 432	97 067	02 933	86 366	84 229	98 585	01 415	85 645	56
5	$\bar{1}$.83 446	$\bar{1}$.97 092	0.02 908	$\bar{1}$.86 354	$\bar{1}$.84 242	$\bar{1}$.98 610	0.01 390	$\bar{1}$.85 632	55
6	83 459	97 118	02 882	86 342	84 255	98 635	01 365	85 620	54
7	83 473	97 143	02 857	86 330	84 269	98 661	01 339	85 608	53
8	83 486	97 168	02 832	86 318	84 282	98 686	01 314	85 596	52
9	83 500	97 193	02 807	86 306	84 295	98 711	01 289	85 583	51
10	$\bar{1}$.83 513	$\bar{1}$.97 219	0.02 781	$\bar{1}$.86 295	$\bar{1}$.84 308	$\bar{1}$.98 737	0.01 263	$\bar{1}$.85 571	50
11	83 527	97 244	02 756	86 283	84 321	98 762	01 238	85 559	49
12	83 540	97 269	02 731	86 271	84 334	98 787	01 213	85 547	48
13	83 554	97 295	02 705	86 259	84 347	98 812	01 188	85 534	47
14	83 567	97 320	02 680	86 247	84 360	98 838	01 162	85 522	46
15	$\bar{1}$.83 581	$\bar{1}$.97 345	0.02 655	$\bar{1}$.86 235	$\bar{1}$.84 373	$\bar{1}$.98 863	0.01 137	$\bar{1}$.85 510	45
16	83 594	97 371	02 629	86 223	84 385	98 888	01 112	85 497	44
17	83 608	97 396	02 604	86 211	84 398	98 913	01 087	85 485	43
18	83 621	97 421	02 579	86 200	84 411	98 939	01 061	85 473	42
19	83 634	97 447	02 553	86 188	84 424	98 964	01 036	85 460	41
20	$\bar{1}$.83 648	$\bar{1}$.97 472	0.02 528	$\bar{1}$.86 176	$\bar{1}$.84 437	$\bar{1}$.98 989	0.01 011	$\bar{1}$.85 448	40
21	83 661	97 497	02 503	86 164	84 450	99 015	00 985	85 436	39
22	83 674	97 523	02 477	86 152	84 463	99 040	00 960	85 423	38
23	83 688	97 548	02 452	86 140	84 476	99 065	00 935	85 411	37
24	83 701	97 573	02 427	86 128	84 489	99 090	00 910	85 399	36
25	$\bar{1}$.83 715	$\bar{1}$.97 598	0.02 402	$\bar{1}$.86 116	$\bar{1}$.84 502	$\bar{1}$.99 116	0.00 884	$\bar{1}$.85 386	35
26	83 728	97 624	02 376	86 104	84 515	99 141	00 859	85 374	34
27	83 741	97 649	02 351	86 092	84 528	99 166	00 834	85 361	33
28	83 755	97 674	02 326	86 080	84 540	99 191	00 809	85 349	32
29	83 768	97 700	02 300	86 068	84 553	99 217	00 783	85 337	31
30	$\bar{1}$.83 781	$\bar{1}$.97 725	0.02 275	$\bar{1}$.86 056	$\bar{1}$.84 566	$\bar{1}$.99 242	0.00 758	$\bar{1}$.85 324	30
31	83 795	97 750	02 250	86 044	84 579	99 267	00 733	85 312	29
32	83 808	97 776	02 224	86 032	84 592	99 293	00 707	85 299	28
33	83 821	97 801	02 199	86 020	84 605	99 318	00 682	85 287	27
34	83 834	97 826	02 174	86 008	84 618	99 343	00 657	85 274	26
35	$\bar{1}$.83 848	$\bar{1}$.97 851	0.02 149	$\bar{1}$.85 996	$\bar{1}$.84 630	$\bar{1}$.99 368	0.00 632	$\bar{1}$.85 262	25
36	83 861	97 877	02 123	85 984	84 643	99 394	00 606	85 250	24
37	83 874	97 902	02 098	85 972	84 656	99 419	00 581	85 237	23
38	83 887	97 927	02 073	85 960	84 669	99 444	00 556	85 225	22
39	83 901	97 953	02 047	85 948	84 682	99 469	00 531	85 212	21
40	$\bar{1}$.83 914	$\bar{1}$.97 978	0.02 022	$\bar{1}$.85 936	$\bar{1}$.84 694	$\bar{1}$.99 495	0.00 505	$\bar{1}$.85 200	20
41	83 927	98 003	01 997	85 924	84 707	99 520	00 480	85 187	19
42	83 940	98 029	01 971	85 912	84 720	99 545	00 455	85 175	18
43	83 954	98 054	01 946	85 900	84 733	99 570	00 430	85 162	17
44	83 967	98 079	01 921	85 888	84 745	99 596	00 404	85 150	16
45	$\bar{1}$.83 980	$\bar{1}$.98 104	0.01 896	$\bar{1}$.85 876	$\bar{1}$.84 758	$\bar{1}$.99 621	0.00 379	$\bar{1}$.85 137	15
46	83 993	98 130	01 870	85 864	84 771	99 646	00 354	85 125	14
47	84 006	98 155	01 845	85 851	84 784	99 672	00 328	85 112	13
48	84 020	98 180	01 820	85 839	84 796	99 697	00 303	85 100	12
49	84 033	98 206	01 794	85 827	84 809	99 722	00 278	85 087	11
50	$\bar{1}$.84 046	$\bar{1}$.98 231	0.01 769	$\bar{1}$.85 815	$\bar{1}$.84 822	$\bar{1}$.99 747	0.00 253	$\bar{1}$.85 074	10
51	84 059	98 256	01 744	85 803	84 835	99 773	00 227	85 062	9
52	84 072	98 281	01 719	85 791	84 847	99 798	00 202	85 049	8
53	84 085	98 307	01 693	85 779	84 860	99 823	00 177	85 037	7
54	84 098	98 332	01 668	85 766	84 873	99 848	00 152	85 024	6
55	$\bar{1}$.84 112	$\bar{1}$.98 357	0.01 643	$\bar{1}$.85 754	$\bar{1}$.84 885	$\bar{1}$.99 874	0.00 126	$\bar{1}$.85 012	5
56	84 125	98 383	01 617	85 742	84 898	99 899	00 101	84 999	4
57	84 138	98 408	01 592	85 730	84 911	99 924	00 076	84 986	3
58	84 151	98 433	01 567	85 718	84 923	99 949	00 051	84 974	2
59	84 164	98 458	01 542	85 706	84 936	$\bar{1}$.99 975	00 025	84 961	1
60′	$\bar{1}$.84 177	$\bar{1}$.98 484	0.01 516	$\bar{1}$.85 693	$\bar{1}$.84 949	0.00 000	0.00 000	$\bar{1}$.84 949	0′
	log cos	log cot	log tan	log sin	log cos	log cot	log tan	log sin	′
	46°				45°				

CONSTANTS.

MATHEMATICAL CONSTANTS.

Quantity.	Numerical Value.		Common Logarithm.
	2.71 828 18	Base of Napierian, natural, or hyperbolic logarithms.	0.43 429 45
$1/\log_{10}e$	2.30 258 5	Factor to multiply into common logs to convert into Napierian logs.	0.36 221 57
$\log_{10}e$	0.43 429 45	Factor to multiply into Napierian logs to convert into common logs.	$\bar{1}$.63 778 43
π	3.14 159 265	Ratio of circumference to diameter.	0.49 714 99
π^2	9.86 960 44	Square of π.	0.99 429 97
1 radian	57° 17′ 45″	57.°29 58 = 206 265.″ = arc equal to radius.	

UNITED STATES, BRITISH, AND METRIC UNITS.

Note.—The following ratios are given on the authority of the U. S. Coast and Geodetic Survey, "Tables of Weights and Measures. Washington, D. C., 1890."

1 metre	39.37 inches.	This is the legalized ratio for the U. S. The U. S. and the British inch are equal. By comparisons to date (July, 1895), it appears probable that this value is smaller than the real ratio of the "Metre des Archives" to the thirty-sixth part of the "Imperial Standard Yard" by one or two parts in one million.	1.59 516 54
1 metre	1.09 361 1 yard.	The U. S. and the British yard are equal.	0.03 886 29
1 metre	3.28 08 33 feet.		0.51 598 42
1 kilometer	0.62 13 70 mile	of 5280 feet.	$\bar{1}$.79 335 03
1 mile	1.60 934 7 kilom.		0.20 664 97
1 yard	0.91 440 2 metre.		$\bar{1}$.96 113 71
1 foot	0.30 480 1 metre.		$\bar{1}$.48 401 58
1 inch	25.40 00$\bar{5}$ mm.	Deduced from above legalized ratio of yard and metre in U. S.	1.40 483 46
1 inch	25.40 000 mm.	is more convenient besides being probably more exact. It is probably about one part in one million too small, as the reciprocal of 0.0254 is 39.37 008.	1.40 483 37

CONSTANTS.

Quantity.	Numerical Value.		Common Logarithm.
1 pound Av.	7000 grains.	The pound avoirdupois is the same in Great Britain and the U. S.	3.84 509 80
1 pound Av.	453.59 242 77 grammes.		2.65 666 58
1 ounce Av.	28.34 953 grammes.		1.45 254 59
1 ounce Troy	31.10 348 grammes.		1.49 280 91
1 grain	0.06 479 892 gramme.	Avoirdupois and Troy grains are the same.	$\bar{2}$.81 156 78
1 kilogramme	2.20 462 2 pounds Av.		0.34 333 42
1 gramme	15.43 235 639 grains.		1.18 843 22
1 litre	1.05 668 U. S. quarts.	By original definition one litre was the volume of one cubic decimetre, but at present the accepted definition is that provisionally adopted by the International Bureau of Weights and Measures in 1880, viz.: the volume of one kilogramme of water at its maximum density. The experimental determination with high accuracy of the relation between this volume and the cubic decimetre is still unfinished. The following values assume this ratio to be unity.	0.02 394 4
1 litre	0.26 417 U. S. gallon.		$\bar{1}$.42 188 4
1 litre	33.814 U. S. fluid oz.		1.52 910
1 quart, U. S.	0.94 636 litre.		$\bar{1}$.97 605 6
1 gallon, U. S.	3.78 544 litres.		0.57 811 6
1 fluid ounce	0.02 957 3 litre.		$\bar{2}$.47 090
1 gallon, U. S.	231 cu. inches.		2.36 361 20
1 British gallon	4.54 346 litres.		0.65 738 67
1 British bushel	36.34 77 litres.		1.56 047 69

MECHANICAL OR DYNAMICAL EQUIVALENT OF HEAT.

The best values of this quantity (usually denoted by J) at present attainable (November, 1895) are the following. The values are uncertain by only about ± one twentieth of one per cent.

427.3 kilogrammetres of work or energy are required at latitude 45°, sea-level ($g = 980.6$ c.g.s.), to raise 1 kilogramme of water through 1° C. at 15° C.

778.8 ft. lbs. of work or energy are required at latitude 45°, sea-level ($g = 980.6$ c.g.s.), to raise 1 lb. of water through 1° Fahr. at 59° Fahr. (= 15° C.). For most engineering purposes 779 ft. lbs. would be near enough.

CONSTANTS.

1402 ft. lbs. of work or energy are required at latitude 45°, sea-level ($g = 980.6$), to raise 1 lb. of water through 1° Cent. at 15° C. For most engineering purposes 1400 ft. lbs. would be near enough.

4.190·10⁷ ergs of work or energy are required to raise 1 gramme of water through 1° C. at 15° C.

To reduce these values to any given locality, multiply by the ratio $g_{45} : g$, where g_{45} is the value (980.6) of the acceleration of gravity at latitude 45°, sea-level, and g is the value at the given place. The latter may be obtained from the latitude and altitude of the place by the formula given upon the next page, unless otherwise better known. The altitude correction is but six one-thousandths of one per cent (0.00 006) for each 1000 ft. of elevation, and therefore quite negligible. Within the limits of uncertainty of the quantities involved the latitude correction for places between 30° and 60° may be applied thus: —

	427.3	778.8	1402
For each degree of latitude **north** of 45° **subtract**	0.04 kgm.	0.07 ft. lbs.	0.13 ft. lbs.
For each degree of latitude **south** of 45° **add**	0.04 kgm.	0.07 ft. lbs.	0.13 ft. lbs.

NOTE. — The persistence with which the time-honored values, 772 ft. lbs. and 424 kgm., of this most important constant are adhered to in practice, although known to be nearly one per cent too small, is due largely to the flagrant negligence of the authors of text-books of both physics and engineering. No attention is paid to the fact that Joule's original data have been amended *acceptably to him*, and that his work has been supplemented by the elaborate researches of at least three other independent observers with radically diverse methods. How remarkably these new results check each other and confirm Joule's *amended* results may be seen from the following table, which is given to indicate the source of the foregoing values.

Authority.	Original Data.			Reduced to Lat. 45° Sea-Level.	Diffs. from Mean.	Date.	Reference.
	J. kgm.	g.	t°.C.				
Joule (as corrected and reduced to Baltimore by Rowland.) [Assigning eq. wts. to all methods.]	427.99	980.05	15.°				See quotations in the Rowland and Griffiths references.
[Assigning Rowland's arbitrary wts.]	426.66	980.05	15.°				
Mean of both.	427.33	980.05	15.°	427.08	—.16	1847–78	
Rowland [at Baltimore].	427.4	980.05	15.°	427.16	—.08	1879	Proc. Am. Acad. A. and S. xv. 75 (1880).
Griffiths [at Greenwich]. (In ergs.)	4.194.10⁷	981.17	15.°	427.70	+.46	1892	Phil. Transac. clxxxiv. 496 (1893).
Miculescu [at Paris].	426.84	981.00	15.°	427.01	—.23	1892	Ann. de Ch. et de Ph. xxvii. 237 (1892).
Mean of all.				427.24	±.23		
Mean omitting Joule.				427.29			

The specific heat of water, and therefore the value of J, diminishes slightly with rise of temperature. The rate of this diminution is not yet satisfactorily determined, but about as nearly as it is now known the true specific heat s_t at any temperature $t°$ not far from 15° C., may be expressed in terms of true specific heat s_{15} at 15° C. by

$$s_t = s_{15}\,[1. - 0.00\,030\,(t - 15)].$$

Hence J_t, the number of kgm. or ergs necessary to raise 1 kgr. of water from $t°$ to $t° + 1°$, will be

$$J_t = J_{15}\,[1. - 0.00\,030\,(t - 15)] \qquad [\text{Range } 13° - 20°],$$

or for 1 lb. of water 1° Fahr.

$$J_t = J_{59}[1 - 0.00030(t - 59)] \qquad [\text{Range } 56^\circ - 68^\circ].$$

The values of J_{15} and J_{59} are given on pages 71 and 72. For further discussion of this subject consult GRIFFITHS, Phil. Mag. xi. 431 (1895).

The scale of temperature in which these results are expressed is the hydrogen scale of the International Bureau of Weights and Measures, which represents, as nearly as it is known, the Thomson absolute scale.

VALUE OF g AT DIFFERENT LATITUDES AND ELEVATIONS.

$$g = g_{45,0}(1 - 0.00\,259 \cos 2\lambda - 0.00\,000\,020\ H).$$

$g_{45,0} = 980.6 \dfrac{\text{cm}}{\text{sec}^2}$ approx. This is the approximate average value of g at latitude 45°, sea-level. The experimental values vary widely in the next place of figures.

λ = latitude of place.

H = altitude of place above sea-level in metres.

NOTE. — Very recent observations render it probable that near the earth's surface the coefficient of H is more nearly 0.00 000 030.

BROWN AND SHARPE WIRE GAUGE.

The diameter corresponding to any gauge number above zero (*i.e.* of any size less than that of a No. 0 wire) may be found to within one hundred-thousandth of an inch (five decimal places) by the expression

$$\left.\begin{array}{r}\text{Diam. in inches of}\\ \text{gauge number } n\end{array}\right\} = 0.32\,486 \times 0.89\,052\,5^n,$$

or, $\quad$ Log. of diam. in inches $= \bar{1}.51\,170 + \bar{1}.94\,964\,5\,n,$

or, $\quad$ " " " " " $= 9.51\,170 - 10. + (9.94\,964\,5 - 10.)n.$

The diameter corresponding to any gauge numbers 0, 00, 000, and so on, may similarly be computed by the following expressions in which N is the number of zeros.

$$\text{Diam. in inches} = 0.28\,930 \times 1.12\,293^N,$$

or, $\quad$ Log of diam. in inches $= \bar{1}.46\,134\,8 + 0.05\,035\,3\,N.$

The two primary sizes on which the gauge is based are No. 0000, diameter 0.46 inch exactly, and No. 36, diameter 0.005 inch exactly.

PHYSICAL AND CHEMICAL CONSTANTS.

For very reliable and extended tables consult LANDOLT und BÖRNSTEIN, Physikalisch — Chemische Tabellen.

www.ingramcontent.com/pod-product-compliance
Lightning Source LLC
Chambersburg PA
CBHW020313220326
41519CB00067B/1085

* 9 7 8 3 3 3 7 1 3 0 6 1 9 *